Also by Tom Cooper in the Africa@War series:
Great Lakes Holocaust: The First Congo War, 1966–1997 (Volume 13)
Great Lakes Conflagration: The Second Congo War, 1998–2003 (Volume 14)
Libyan Air Wars Part 1 1973-85 (with Albert Grandolini & Arnaud Delalande) (Volume 19).

Co-published in 2015 by:

Helion & Company Limited
26 Willow Road
Solihull
West Midlands
B91 1UE
England
Tel. 0121 705 3393
Fax 0121 711 4075
email: info@helion.co.uk
website: www.helion.co.uk

and

30° South Publishers (Pty) Ltd.
16 Ivy Road
Pinetown 3610
South Africa
email: info@30degreessouth.co.za
website: www.30degreessouth.co.za

Designed & typeset by Farr out Publications, Wokingham, Berkshire
Cover design by Paul Hewitt, Battlefield Design (www.battlefield-design.co.uk)

Printed for Helion & Co by Henry Ling Ltd., Dorchester, Dorset and for 30° South Publishers by Pinetown Printers, Durban, South Africa

ISBN 978-1-909982-38-3

British Library Cataloguing-in-Publication Data
A catalogue record for this book is available from the British Library

Front cover: This photograph of a group of Ethiopian pilots – led by Negussie Zergaw (first from left) – in front of a recently-delivered MiG-21bis, is nicelly illustrating the story of their air force during the Ogaden War: nearly all of them were trained in the USA and used to fly T-33s, F-86Fs, and then F-5s, before being sent to the USSR for conversion training on MiG-21s. (EtAF)

CONTENTS

Note: In order to simplify the use of this book, all names, locations and geographic designations are as provided in *The Times World Atlas*, or other traditionally accepted major sources of reference, as of the time of described events.

Abbreviations

AAA	Anti-aircraft artillery
AA-2 Atoll	ASCC code for R-3S or R-13, Soviet air-to-air missile
AB	Air base
AdA	Armée de l'Air (French Air Force)
AK	Automat Kalashnikova; general designation for a class of Russian or former Eastern Bloc manufactured 7.62mm assault rifles
An	Antonov (the design bureau led by Oleg Antonov)
APC	Armoured personnel carrier
ASCC	Air Standardisation Co-ordinating Committee (US, British, Australian and New Zealand committee for standardisation of designations of foreign aircraft)
Brig Gen	Brigadier General (military commissioned officer rank)
CAP	Combat air patrol
Capt	Captain (military commissioned officer rank)
CAS	Close air support
CAdS	Corpo Aeronautico della Somalia (Somali Aeronautical Corps)
CBU	Cluster bomb unit
CCS	Ciidanka Cirka Soomaliaayed (Somali Air Force)
C-in-C	Commander-in-Chief
CO	Commanding Officer
COIN	Counterinsurgency
Col	Colonel (military commissioned officer rank)
Col Gen	Colonel General (top military commissioned officer rank)
CPSU	Communist Party of the Soviet Union
DAAFAR	Defesa Anti-Aérea y Fuerza Aérea Revolucionaria (Cuban Air Defence Force & Air Force; often shortened to 'FAR' in every-day conversation)
DoD	Department of Defence (USA)
EAL	Ethiopian Airlines
EE	English Electric
ELA	Eritrean Liberation Army
ELF	Eritrean Liberation Front
ELINT	Electronic intelligence
EPLF	Eritrean People's Liberation Forces
EPRDF	Ethiopian People's Revolutionary Democratic Front
EtAF	Ethiopian Air Force
Flt Off	Flight Officer (military commissioned officer rank)
Flt Lt	Flight Lieutenant (military commissioned officer rank, equal to Captain)
FS	Fighter squadron
GCI	Ground controlled interception
Gen	General (military commissioned officer rank)
GP	General-purpose (bomb)
HE	High explosive
HQ	Headquarters
IAI	Israeli Aircraft Industries (since 2006 Israeli Aerospace Industries)
IAP	International airport
IDF	Israeli Defence Force
IEA	Imperial Ethiopian Aviation
IEAF	Imperial Ethiopian Air Force
IEAA	Imperial Ethiopian Army Aviation
IFF	Identification friend or foe
IFV	Infantry fighting vehicle
IR	Infra-red, electromagnetic radiation heat sensor
Il	Ilyushin (the design bureau led by Sergey Vladimirovich Ilyushin, also known as OKB-39)
IIAF	Imperial Iranian Air Force
IRIAF	Islamic Republic of Iran Air Force
KGB	Komitet Gosudarstvennoy Bezopasnosti (Committee for State Security. Soviet National Security Agency, 1954–1991)
KIA	Killed in action
km	Kilometre
Lt	Lieutenant (military commissioned officer rank)
Lt Col	Lieutenant Colonel (military commissioned officer rank)
1st Lt	First Lieutenant (military commissioned officer rank)
2nd Lt	Second Lieutenant (lowest military commissioned officer rank)
MAAG	Military Aid and Assistance Group
Maj	Major (military commissioned officer rank)
Maj Gen	Major General (military commissioned officer rank)
MANPAD(S)	Man-portable air defence system(s) – light surface-to-air missile system that can be carried and deployed in combat by a single soldier
MBT	Main Battle Tank
Mi	Mil (Soviet/Russian helicopter designer and manufacturer)
MIA	Missing in action
MiG	Mikoyan i Gurevich (the design bureau led by Artyom Ivanovich Mikoyan and Mikhail Iosifovich Gurevich, also known as OKB-155 or MMZ 'Zenit')
MoD	Ministry of Defence
MRL	Multiple rocket launcher
NCO	Non-commissioned officer
OAU	Organisation of African Unity
PDRYAF	People's Democratic Republic of Yemen Air Force (air force of former South Yemen)
PLO	Palestinian Liberation Organisation
PMAC	Provisional Military Administrative Committee (120-member committee of Ethiopian officers, better known as the 'Derg' or 'Dergue')
PoW	Prisoner of War
R-3S	Soviet-made air-to-air missile (see AA-2)
RA	Regia Aeronautica (Royal Italian Air Force)
RAF	Royal Air Force
RHAW	Radar homing and warning system
RWR	Radar Warning Receiver
SA-2 Guideline	ASCC code for S-75 Dvina, Soviet SAM system
SA-3 Goa	ASCC code for S-125 Neva, Soviet SAM

	system
SA-7 Grail	ASCC code for 9K32 Strela-2, Soviet MANPAD
SA-13	ASCC code for Strela-10, Soviet SAM system
SAAB	Svenska Aeroplan Aktiebolaget (Swedish aircraft manufacturer)
SAC	Somali Aeronautical Corps
SALF	Somali-Abo Liberation Front
SAM	Surface-to-air missile
SMSC	Supreme Military Strategic Committee
SNA	Somali National Army
SNDF	Somali National Defence Force
Sqn	Squadron
Sqn Ldr	Squadron Leader (military commissioned officer rank, equal to major)
UN	United Nations
UNHCR	UN High Commissioner for Refugees
UNRWA	United Nations Relief and Works Agency
USAAF	United States Army Air Force (until 1947)
USAF	United States Air Force (since 1947)
US$	United States Dollar
USSR	Union of Soviet Socialist Republics
V-TA	Komandovaniye voyenno-transportnoy aviatsii (Soviet Military Transport Aviation)
V-VS	Voyenno-Vozdushnye Sily (Soviet Air Force)
WIA	Wounded in action
WSLF	Western Somali Liberation Front
WWI	First World War
WWII	Second World War

CHAPTER 1
GEOPOLITICAL PRELUDE

In the late 1970s, as the Cold War between the US-led North Atlantic Treaty Organisation (NATO) and communist states of Central and Eastern Europe grouped within the Soviet dominated Warsaw Pact was heading for its next high point, rumours about an intensive conventional war between Ethiopia and Somalia began spreading through the circles of various military intelligence agencies and academies around the world. The word was that Somalia (usually described as a Soviet 'client state' until that time) had invaded Ethiopia (formerly a major US ally in Africa) in order to realise its national aim of annexation of the area called 'Ogaden'. The two nations then swapped their alliances to the superpowers in the middle of the resulting war, with Ethiopia joining the Soviet bloc and Somalia turning West. Both sides intensively deployed Soviet-built fighter-bombers of Mikoyan i Gurevich (MiG) design. Cuban advisers became involved on the Ethiopian side, followed by the Soviets and even 'Israeli mercenaries' that flew for the Ethiopian Air Force (EtAF). A 'Soviet General' then launched a daring heliborne operation to outflank and rout the Somalis and force them out of Ogaden, thus successfully concluding the war in Ethiopia's favour.

For years after the end of this conflict very little factual information was available. Although a number of publications related to labyrinthine political and diplomatic intrigues that surround all aspects of its history were released, none offered any closer insights into specifics of combat operations. Contrary to earlier reports, it appeared that this conflict did not bring any new, interesting or relevant experiences, and especially that – with exception of one operation – air power did not play any kind of important role. Indeed, in a classic misjudgement about this war, a conclusion emerged that US-built Northrop F-5 Tiger IIs, which were known to have been delivered to Ethiopia in the mid-1970s, and Soviet-built MiG-21s, known to have been operated by Somalia around the same time – two of most-prolific fighter-bomber types in service with more than 70 different air forces around the world – have never engaged in air combat, and that thus it is only possible to 'guess' about the outcome of any clashes between them.

It was only in the early 2000s – and thanks to the ever wider availability of the internet – that additional, more authoritative information began to appear. Gradually, a picture emerged according to which the Somali invasion of Ogaden ignited a major war that initially brought the Somalis close to attaining their strategic goal but eventually turned into a battle of attrition that Somalia could not sustain. Indeed, with extensive support from Cuba and the Soviet Union, the Ethiopians not only averted catastrophe but also turned the tables on the aggressor. Even more time lapsed before it transpired that in opinion of many Ethiopians, their country was saved by its small, but well-equipped, organised, and trained air force, and a small number of Northrop F-5E Tiger II fighter-bombers almost in the same manner the RAF saved the United Kingdom during the Battle of Britain, in 1940.

A result of years of often troublesome research, put together with help and information from very different sources, this book is probably the first to cover in great depth the emergence of the Ethiopian and Somali air forces, and the air war over Ogaden against the backdrop of ground warfare. Unfortunately, it is still likely to contain some errors and omissions. Sadly, official Ethiopian archives remain well outside the reach of most researchers and practically all Somali documentation was destroyed in the course of wars that savaged this country over the last 30 years.[1] Except for a handful of rarely available books in Amharic language, there is also next to no authoritative literature about the armed forces of either nation, while the few articles published in specialised magazines over the years mainly emphasised the work of foreign instructors in the 1950s and 1960s. Even with the help of several participants from both sides, some details prefer to remain unknown: memories fade with years, no matter how many have their recollections and images firmly blazed into their minds.

Geography

Situated at the north-eastern corner of Africa, close to the Arabian Peninsula and on the coast of the Indian Ocean, the Gulf of Aden

1 The only researcher ever granted official permission to work in the official Ethiopian archives so far was Gebru Tareke, author of the essay 'The Ethiopia–Somalia War', published in 2000 in *The International Journal of African Historical Studies*. Tareke's work mainly covers ground warfare, but is considered the most complete, objective and balanced narrative by all of participants interviewed by the author.

and the Red Sea, the Horn of Africa contains ports and airfields that connect routes of intercontinental significance and strategic importance.

Stretching from Sudan in the north and west, southwards to Kenya and to Djibouti in the east, Ethiopia is the biggest country in this area. More than half of Ethiopian territory is covered by the Ethiopian Plateau, diagonally split in a north-eastern to south-eastern direction by the Great Rift Valley and with an average elevation of about 1,680m above the sea level. The plateau is cut by many rivers and deep valleys, ranging from the Dallow Depression, 100m below sea level to the South Mountains of the central highlands that rise up to 4,000m. Southern Ethiopia is bisected by the 40–60km wide Rift Valley. The road network remains underdeveloped and in many areas of Ogaden is actually non-existent, significantly impeding vehicular movement during rainy seasons. The principal rainy season occurs between mid-June and September, followed by a dry season that may be interrupted in February or March by a short rainy season.

Although largely homogenous in regards of religion, the population of Ethiopia is deeply divided along ethnic, regional, and political lines, and for centuries the country has faced an uphill struggle to keep all of these united. The Amhara, who founded the original nation, and the related Tigreans, both of which are highland peoples of partly Semitic origin, constitute around 30% of the total population. They primarily occupy the north-western Ethiopian highlands and the area north of Addis Ababa. Central and south-western Ethiopia is largely populated by the Oromo, a pastoral and agricultural people that constitute up to 40% of the population. Western Ethiopia is predominantly populated by the Shankella, that constitute about 6% of the population, while the east and south-east is predominantly populated by the Somali. Out of around 70 languages spoken in Ethiopia, most belong to the Semitic and Cushitic branches of Afro-Asiatic family. Amharic, the official language, is spoken by more than half of the population, but English and Arabic are also widely spoken.

The country is divided into nine regions that have a significant degree of autonomy and are composed around specific ethnic groups: Tigray, Amhara, Afar, Oromia, Somalia, Benishangul-Gumaz, Gambela, Harari, and Southern Nations, Nationalities and Peoples' which comprise about 41 different ethnic groups. Addis Ababa is the largest city in Ethiopia, but only about 15–20% of the population can be classified as urban.

About 40% of Ethiopians are Christians, primarily followers of the Ethiopian Orthodox Union church, an autonomous Christian sect headed by a patriarch and closely related to the Coptic church of Egypt (which was the state church of Ethiopia until 1974). All the southern and eastern regions have Muslim majorities, who represent about 45% of the country's population. The south also contains considerable numbers of animists. The majority of members of the sect known as Beta Israel or Falashas, who practiced a type of Judaism that probably dates back to contact with early Arabian Jews, were airlifted to Israel in 1991.

The economy is heavily dependent on earnings from the agricultural sector, with the raising of livestock being the most characteristic form of economic activity, followed by farming coffee, cotton, sugar, fruit and vegetables. Much of the trading is conducted by barter in local markets, especially because periodic droughts have greatly reduced agricultural output and repeatedly forced the country to import basic foodstuffs.

Somalia covers most of the coastline of the Indian Ocean and the Gulf of Aden, stretching over 3,025km (1,880 miles) from Djibouti in the north, to Kenya in the south, and is bordered by Ethiopia in the west. The terrain in this country is dominated by dry savannah plains characterised by lava rocks and sand that pose formidable obstacles to wheeled vehicles in several areas. A series of mountain ranges, with average elevations between about 900 and 2,000 metres (about 3,000 and 7,000ft), dominate the northern part of the country, while a sandy coastal plain borders the Gulf of Aden in the north and a wide coastal plain with many sand dunes borders on the Indian Ocean in the south. The country's two major rivers, the Jubba (or Genale) and the Shabele (or Shebelle), are found on the southern plateau and there are very few natural harbours. Like that of Ethiopia, the climate of Somalia ranges from tropical to subtropical and from arid to semiarid. But, the monsoon wind brings a dry season from September to December and a rainy season from March to May.

While nominally divided into 18 regions and 84 districts, Somalia is one of very few African states where virtually all citizens share a history, language, culture and religion. The vast majority of the population consists of Somali, a Cushitic people, some 70% of which are nomadic or semi-nomadic pastoralists. The reminder, including Arabs, Bantu-speaking people in the southern part of the country, some Indians, Italians and Pakistanis, are either crop farmers or inhabitants of the few urban centres. Islam is the state religion and most of the people are Sunni Muslims. The official language is Somali, but Arabic, English and Italian are also spoken.

Between the Ethiopian Plain and Somalian Plateau lies the area of Ogaden. Some 200,000 square kilometres (125,000 square miles) in size, this barren and bleak region is drained by the Shabele and Juba rivers, but otherwise predominated by a semi-desert of sandstone and limestone, with some coarse grass and a few stunted thorn and acacia trees. Flat-topped hills and arid plains slop southward from the Harar Plateau (elevation 2,000m, or around 6,000ft), where aromatic flora, producing frankincense and myrrh, are indigenous. As of the mid-1970s, it was inhabited almost exclusively by some 500,000 ethnic Somalis, mainly nomadic herdsmen of the Ogaden clan, which gave the territory its name. While mineral resources are relatively diverse and include deposits of petroleum, copper, manganese and uranium, they have not been exploited. Instead, the economy in this part of Ethiopia since the 16th century has been dominated by Somali nomadic pastoralists that graze their herds on the plains. The climate is generally hot and dry throughout most of the year.

Ethiopian Heritage

Ogaden has long been a site of contention: first between Christian Abyssinia and the Muslim emirs, then between Ethiopia and European colonial states, and finally between Ethiopia and the Somali nation.

Ethiopia is a country situated in one of the oldest – if not *the* oldest – area of human habitation. Archaeological research has shown that modern Homo sapiens probably evolved there. The original form of the modern-day name of this country was first used by ancient Greeks to refer to the peoples living south of ancient Egypt; modern usage has transferred this name further south, to the land of people known until the early 20th century as 'Abyssinia'.

In the 1st century AD, the Axumite Empire developed in the area. Relatively isolated due to the inaccessibility of the high central plateau, rich with gold, iron and salt deposits, it eventually became one of the five largest empires of the world in its time. It was in the year AD 330 that it experienced the introduction of Christianity through Greek clergy. Through the following two centuries, the

Axumite Empire benefited from a major transformation of the maritime trading system that linked the Roman Empire and India. This increased the significance of the Red Sea as a maritime trading route that made Axum's main port, Adulis, a major trading centre. At its height, the Axumite Empire controlled the area covering the entire modern-day Ethiopia, Eritrea, northern Sudan, southern Egypt, Djibouti, Yemen and southern Arabia.

Axum remained strong until the rise of Islam, in the 7th century. Because the Axumites had sheltered Muhammad's first followers, the Muslims never attempted to conquer the country as they spread across Africa. While Axumite naval power gradually declined through that period, in AD 702 its pirates were able to invade the Hejaz and occupy Jeddah. In retaliation, the Muslims took the Dahlak Archipelago from Axum and began spreading along the coast of the Red Sea, forcing Axum into isolation from the rest of the world.

In medieval times, three chief provinces came into being: Tigray in the north, Amhara in the centre and Shewa in the south. The seat of the government was usually in Amhara, but at times there were two or even three kings reigning at the same time. It was only in 1528 that Ethiopia again made contact with the outside world. Invaded by a Muslim army from the nearby Sultanate of Adal, the Negus ('King') Lebna Dengle Dawit II requested help from Portugal. As Muslim forces came close to extinguishing the ancient realm of Ethiopia and converting all of its subjects to Islam, the Portuguese expedition led by Cristóvão da Gama arrived in 1541 and saved the nation. However, they were subsequently obliged to make their way out of Ethiopia and the area that is now in Somalia.

Many historians trace the origins of hostility between Ethiopia and Somalia to this war, but the conflict of 1528–1541 also resulted in bitter religious conflicts with the Jesuits and inner struggles between different Ethiopian rulers and the country remained relatively isolated for the following 300 years. It was not before 1855, when Lij Kassa proclaimed himself 'Negus Negusti' ('King of Kings') under the name of Tewodros II and launched a campaign to unite the nation under his rule, that modernisation and the opening up of Ethiopia began. Although a ruthless ruler, Tewodros was determined to protect the country from the Europeans who were scrambling to get colonies in Africa at that time. When Queen Victoria failed to answer his letter, in 1867, he took it as an insult and imprisoned several British residents, including the consul. The British deployed an army of 12,000 from Mumbai to Ethiopia and defeated Tewodros during the battle at Magdala (better known as Amba Mariam), prompting him to commit suicide.

Effects of the Suez Canal

The end of Tewodros' rule resulted in an internal power struggle, won by Kassa, who was crowned Emperor Yohannes IV and rose to power at the time the area of the Red Sea became strategically important due to the opening of the Suez Canal. As Western colonial nations began political battles for the control over the shores, the British occupied Yemen, the French took Obock, Asars and Issa, while the Ethiopians had the ambition to conquer the source of the Nile and had invaded Sudan. In 1870, the Italians appeared on the scene, buying the port of Asseb from the local sultan. In 1888, the Italians exploited Yohannes IV's preoccupation with defending Ethiopia from an invasion of dervishes from Sudan and deployed 20,000 troops in the country. Not interested in fighting the newcomers, the Emperor solved all the disputes – more or less – through negotiations, and granted permission for some 5,000 troops to remain stationed in a part of the Ethiopian Tigray Province, which over time became known as 'Eritrea'.

Meanwhile, on 9 March 1889, Yohannes IV had defeated the Dervish invasion, but a stray bullet hit him and his army withdrew. The Emperor died during the night and his body fell into the hands of the enemy. As soon as this news reached Sahle Maryam of Shewa, he proclaimed himself Emperor Menelik II of Ethiopia. Only two months later, Menelik II signed a treaty with Rome, granting Eritrea to Italy in exchange for supply of 30,000 rifles, ammunition and several cannons. The Italians scrambled to declare this treaty as granting them a protectorate over all of Ethiopia. Menelik's protests were completely ignored and this caused another war. The following conflict between Ethiopia and Italy culminated in a humiliating Italian defeat during the battle of Adwa. On 1 March 1896, a provisional peace treaty was concluded in Addis Ababa in which Rome recognised the absolute independence of Ethiopia, which thus became the first internationally recognised independent African state.[2]

Following this success, the Ethiopians invested heavily in development of modern infrastructure, including the construction of Addis Ababa–Djibouti railroad, post and telephone services.[3] The Emperor began appointing ministers, a bank was founded and the first hotel, hospitals and schools opened in the capital.

Menelik died in December 1913 and was succeeded by his grandson, who proved unpopular due to ties to Muslims; he ruled only for three years. In 1916, he was deposed by the Christian nobility who made Menelik's daughter Zauditu, an Empress, with her cousin Ras Tafari Mekonnen (son of a hero of the Battle of Adwa) a regent and successor to the throne. After the death of Empress Zauditu, in 1930, Mekonnen founded his own army and established himself in power after a civil war against different opponents, resulting in his crowning as Emperor Haile Selassie I of Ethiopia.

Italian Revenge

Aiming to expand its colonial possessions, Italy invaded Ethiopia with troops deployed in Eritrea and Italian Somaliland in 1935. The initial advance was slow, but Adwa – the site of the Italian defeat in 1896 – fell on 6 October and the commander of the Italian forces, Gen Emilio De Bono, subsequently launched a major campaign into Tigray, characterised by deployment of air power, tanks and chemical weapons. Although many landlords offered no resistance (some even sided with conquerors) the Italian invasion was no 'walk-over'. Indeed, De Bono's troops were halted by December and he was replaced by Gen Pietro Badoglio.

In an attempt to exploit the apparent Italian weakness, Selassie ordered a counter offensive, only to have his forces battered by the full might of heavy and chemical weapons. Putting to the test the tactics of Gen Giulio Douhet, the invaders then launched a bitter campaign of air attacks and artillery bombardments, hitting not only the Ethiopian Army's positions but also civilian settlements. The town of Harar was fire-bombed on 29 March 1936, and two days later Italy won a decisive battle at Maychew. Addis Ababa fell on 5 May and Rome then officially annexed all of Ethiopia, together with Eritrea and Somaliland, declaring them elements of the Italian East Africa.

2 Addis Ababa was established as the Ethiopian capital in 1886, in Intoto Valley, in the course of Menelik IV's attempts to re-unite the country, see Marcus, *A History of Ethiopia*, pp. 104–116.

3 Ibid. The concession for a railway was issued to a French company in 1894; construction of the stretch from Djibouti to Dire Dawa, a town some 45km from Harar, was completed on 31 December 1902.

Emperor Selassie with some of the British Army officers during the campaign that led to liberation of Ethiopia form Italian rule in 1941. (Mark Lepko Collection)

Somalis are proud of their military heritage, much of which dates from their defence against the Portuguese in the 16th century and revolts against British and Italian colonial rule (motivated by the Mahadist uprising in Sudan), in the late 19th and early 20th centuries. This Italian illustration from the 1910s shows the cavalry of so-called 'Marehan Conquerors' that rose against the British in 1914. (Mark Lepko Collection)

The Italians did a lot to improve the local infrastructure, constructing roads and bridges and expanding most of the larger cities. However, they brutally crushed repeated revolts of intellectuals, provoking armed resistance. The start of the end of their rule came in the summer 1940, when Rome launched an offensive against British-held Sudan, followed by another into British Somaliland, which saw a successful capture of Berbera. However, heavy terrain and supply problems, as well as the uprising in Ethiopia, prevented their victory and Italian forces eventually found themselves facing a three-prong British and Commonwealth counteroffensive.

Although forced into exile, Emperor Selassie never stopped requesting help from Great Britain and elsewhere. His pleads to the League of Nations went unanswered, but the situation changed fundamentally once Italy and the United Kingdom found themselves on opposing sides during WWII. Following a series of defeats to Commonwealth forces (primarily South Africans), the last major Italian units in Ethiopia capitulated and Emperor Selassie returned to Addis Ababa on 5 May 1941.

Drawing lessons from this war, as well as from the Royal Air Force's (RAF) involvement in crushing of a rebellion in Tigrai Province in September 1943, the Ethiopian ruler subsequently took care to bring his country into the United Nations as a founding member, greatly expand diplomatic relations and establish a new military, including an air force. Furthermore, he established a dependable tax base to support a dependable military that could help him in case of a new invasion from abroad, as well as problems at home. Although many of the nobles and several provinces battled related reforms and laws, over time, Selassie managed to make peace with the many ethnic, religious and economic factions through appeasement and compromise. This resulted in his country enjoying an unprecedented period of relatively uninterrupted stability and progress. Strengthened with tax-income and foreign aid, for most of the time between the 1940s and 1970s, the Ethiopian ruler was able to spend about 40% of the nation's annual budget on defence and internal security.

Vibrant Somalia

Somalia never existed, as such, prior to its independence in 1960. In ancient times, this area was designated the Punt by Egyptians and the inhabitants referred to as 'Black Berbers'. The Ethiopian Kingdom of Aksum controlled most of the Punt between the 1st and 7th centuries AD, while Arab tribes set up trading posts along the coastline of the Gulf of Aden. The Somali people – mostly Islamised Yemeni refugees – began to migrate into this region in the 13th century. During the 15th, 16th and 17th centuries the Portuguese launched several expeditions into the area, but faced stiff resistance while attempting to establish a presence and eventually withdrew when local warriors were reinforced by the Ottoman Empire, which subsequently established itself in control of northern sultanates. The southern ones accepted the sovereignty of the Sultan of Zanzibar. During the 18th and the early 19th centuries, the Gobroon Dynasty established itself in power in some parts of East Africa, reaching its apex shortly before the region gained strategic importance due to the construction of the Suez Canal. The French bought the port of Obock and surrounding area in 1862, which became known as the 'French Territory of the Afars and the Issas', paving the way for creation of present-day Djibouti. Through the mid-1870s, Egypt occupied some of the towns along the Somali coast. When Egyptian troops withdrew due to the Mahdi Revolt in Sudan, the British, mainly concerned with keeping open the route to India, occupied the territory instead. In 1887, Great Britain proclaimed a British Protectorate in what became known as British Somaliland and in 1905, after the British put down several armed revolts by local Dervishes, the area was placed under the administration of the Colonial Office.

Meanwhile in early 1869, Italians began settling in some of the ports of northern Somalia, and by the early 20th century their influence spread through agreements with local Somali chieftains, Great Britain, Zanzibar and Ethiopia. Taking advantage of the Treaty of London, from 1915 the Italians then spread their control inland from Asseb (nowadays in Eritrea) and in 1936 merged all the territories of Somaliland, Eritrea and Ethiopia into the colonial state of Italian East Africa. It was from there that Italian troops invaded British Somaliland in 1940 and expelled the British. However, just a year later the British took their protectorate back and then continued with the successful liberation of Ethiopia.

The Italian Peace Treaty of 1947 forced Rome to give up all of its possessions in Africa and the Allies took over responsibility for the Horn of Africa. However, because the USA, Great Britain, France and the Union of Soviet Socialist Republics (USSR) failed to reach an agreement, the matter was referred to the General Assembly of the UN in 1948. This promised to give Italian Somaliland independence following a period of ten years as a UN trust territory, merging the former British Protectorate of Somaliland with formerly Italian-ruled parts of Somalia in the process. With Rome accepting the UN terms, on 1 April 1950 the British military government was replaced by a provisional Italian administration and the territory named Somalia. In accordance with the UN Trusteeship Council, the former British Somaliland gained independence on 26 June 1960, and was then united with the former Italian-Somaliland, for combined independence on 1 July 1960. The new Somalia was founded as a parliamentary republic, with Aden Abdullah Osman Daar becoming the first president. However, it did not include predominantly Somali-populated parts of Kenya, nor Affars and Issas (Djibouti).[4]

4 France released Affars and Issas into independence as Djibouti in 1977, but the population decided not to merge their country with Somalia.

CHAPTER 2
MILITARY BACKGROUNDS

Origins of the Imperial Ethiopian Air Force

The history of military flying in Ethiopia can be traced back to November 1922, when (then Ras) Tafari Mekonnen witnessed a show of the RAF in Aden. Having never seen an airplane before, he was captivated by this demonstration and spontaneously asked to have a closer look at one of the biplanes and to take a flight. The result of this experience was his decision to develop the Imperial Ethiopian Aviation (IEA).[5] Further impressed by the exploits of Hubert Fauntleroy Julian, the first African American to obtain a pilot's licence in the USA, Mekonnen visited Europe to negotiate the purchase of some French and German aircraft two years later. These were used to form the nucleus of an air force, based on the race course at Bishoftu, near Addis Ababa, and officially organised into the IEA on 18 August 1929. That was the day when Frenchman Andre Maillet, a former fighter pilot of the Armée de l'Air (AdA, French Air Force) during the First World War (WWI), delivered the first Potez 25A2 biplane to Ethiopia. An additional five French, one Italian and one British built biplanes, were all acquired by November 1930.

Aiming to set up a flying school, another French pilot, Gaston Videl, arrived in Addis Ababa in April 1930. His first students included Mishka Babitcheff (the son of a White Russian and Ethiopian mother), Asfaw Ali, Seyoum Kebed, Bahru Kaba, Demissie Haileyesus, Demke Tekle-Wold, and Mulu-Embet Imru, the first female student pilot in Africa. Two other cadets, Bahru Kabba and Tesfamichael Haile, were sent to France to study flying at Saint Cyr. Asfaw Ali and Mishka Babitscheff graduated first, receiving their pilot certificates from Emperor Haile Selassie I himself, on 13 October 1930. By then, Maillet was succeeded by another Frenchman, Paul Corriger, who served as commander and chief flight instructor of the IEA until the Italian invasion in 1935, when Mishka Babitcheff replaced him.

Meanwhile, new aircraft were purchased in 1934 and 1935, including two Beech B.17 Staggerwings, two Fokkers (one F.VIIa/3m and an F.XVIII) and a single Meindl AVII; the latter became the first aircraft ever assembled in Ethiopia. However, there were still only two qualified Ethiopian pilots, including Mishka Babitcheff and Asfaw Ali. Like the rest of the Ethiopian military, the IEA was still underdeveloped and lacked the ability to defend the country when, on the morning of 3 October 1935, Italy invaded Ethiopia with troops deployed from Italian Somaliland.

Only very few reliable records about IEA operations during the war with Italy are available. It is certain that the Ethiopians had no combat aircraft or combat-trained pilots at the time and stood little chance against the might of Regia Aeronautica (RA, Royal Italian Air Force). Known to have been operational as of 1935 were three Potez 25s, one Junkers W.33, one Breda Ba.15, a de Havilland Tiger Moth D.H.60C, one Beechcraft 17, and a Fiat AS.1. Within days of the

5 Forsgren, SAFO No. 20, p. 23.

Row of six Potez biplanes as seen at Bishoftu in 1930. (EtAF)

Mishka Babitcheff (sitting in cockpit) was one of first Ethiopians to obtain a pilot's licence in 1930. (EtAF)

French instructor Paul Corriger with a group of six Ethiopian cadets and a D.H.60 biplane in 1935. (Photo by Richard Pankhurst)

The two Fokkers acquired by the IEA in 1934 and 1935. (via S. N.)

On their delivery, all early Ethiopian aircraft were painted different civilian colours. Only the French-made Potez biplanes were painted green. This was the third of six Potez 25 A2s operated by the IEA. The plane was nicknamed 'Nesre Mekonnen' (Prince Mekonnen in Amharic), and wore national markings on top wing surfaces (usually in form of a rectangle in green, yellow and red), as well as the Lion of Judah (symbol of Imperial Ethiopia) under the cockpit. (EtAF)

Italian attack, only one Potez 25A and three other aircraft remained operational. Although capable of carrying machine-guns, they were never armed and only used as light transports.

Nevertheless, shortly before and early during the war, several African Americans were recruited or volunteered to go to Ethiopia to serve as professionals in various fields. The most influential of them was Colonel (Col) Julian, assigned the Commander-in-Chief (C-in-C) of the IEA in October 1935. Julian took over as commander of the Flying School and became responsible for the tuition of Mulu-Embebet after she completed her high school education at Lcyee Gebremariam, the French School in Addis Ababa. However, the Italian invasion prevented her further training and she never soloed.

Another US pilot active in Ethiopia at that time was John Robinson, better known as the 'Black' or 'Brown Condor', who earned his wings at the Booker T. Washington's Tuskegee Normal and Industrial Institute, Alabama, in 1920. Invited to Ethiopia by Emperor Selassie, Robinson arrived in late May 1935 and soon became involved in fighting, flying the few operational aircraft, transporting troops, ammunition and supplies, as well as the Emperor, from one site to the other. At some point in time, Julian and Robinson had a public fist-fight, and the former was asked to leave the country. Although frequently confronted by RA fighters and ground fire, and having his aircraft riddled by bullets a number of times, Robinson was wounded only once.

The third US pilot that appeared in Ethiopia during that war was John H. Spencer, who acted as official military adviser. He is known to have flown some transport and reconnaissance missions with one of the Potez 25s, together with the British Military Attaché, Major (Maj) Holt, as well as with Babitscheff.

Initially, the IEA's main operation zone was in the Dessa area, where the Ethiopian Army Field Headquarters (HQ) was also positioned. Dessa included a small airfield that was several times attacked by the RA, and in one instance even the Emperor had to man an anti-aircraft gun. Eventually, all the efforts of the Ethiopian military and US volunteer pilots were in vain and the Italians won the war.

Re-Establishment of Imperial Ethiopian Air Force

Efforts to establish the Imperial Ethiopian Air Force (IEAF) were launched in 1944, when John Robinson returned to Ethiopia, now as a full colonel of the US Army Air Force (USAAF) and together with a team of African-American aviators and technicians. Within weeks he had recruited 30 cadets from schools in Addis Ababa and

One of two Cessna AT-17 Bobcat advanced trainers, donated to Ethiopia by the USA in August 1944. (EtAF)

This was one of two little-known Avro Ansons operated by the IEAF as training aircraft and light bombers during the late 1940s. Both saw only short service in Ethiopia and were returned in 1949. (Albert Grandolini Collection)

One of the first batch of 16 SAAB B.17s delivered to Ethiopia in late 1967. The type was originally designed by the AB Svenska Järnvägsverkstäderna ('Swedish Railway Workshops Company) as a dive-bomber in the 1930s and remained in service with the IEAF until 1977. (EtAF)

While the newly-established IEAF was re-equipped with aircraft of Swedish origin, the Ethiopian Army obtained CKD AH IVb light tanks from Czechoslovakia, armed with two Skoda 7,92mm machine guns. Ordered in June 1948 and delivered by ship to Djibouti and then by rail to Addis Ababa in 1950, these vehicles with a crew of two remained in service until the early 1980s. At least a company of them saw combat service during the Ogaden War. (Albert Grandolini Collection)

Except for US-made aircraft, Ethiopia also acquired 15 M8 Greyhound 6x6 light armoured cars armed with 37mm cannons and two machine guns in 1955, 15 in 1956 and 15 of the M20 variant in 1959. (Albert Grandolini Collection)

Dire Dawa, and acquired two Avro Anson twin-engined training aircraft from Great Britain to form the core of the future air force. Washington then donated two Cessna trainers on 3 August of the same year, and a flying school was established at Harar Meda, with Robinson in command, together with an Egyptian pilot, Capt Sadik. Two de Havilland D.H.60 Tiger Moths, followed by various other light aircraft, were acquired in 1945, and by the next year, the Flying School had 75 students.[6]

However, because neither the USA nor the UK showed interest in seriously expanding the IEAF, and because of quarrels between Robinson and one of Swedes that worked with the air force, Count Carl-Gustav von Rosen, the American was asked to leave in early 1946. Signalling the start of the 'Swedish period' in the history of the Ethiopian Air Force, Selassie then appointed von Rosen as a new chief of the Flying School and the C-in-C of the fledgling IEAF.[7]

Planning to establish a force of three squadrons (including one bomber, one fighter and one combined reconnaissance/bomber unit) and a flight school, von Rosen was soon reinforced by a group of 19 Swedish instructors that arrived in Addis Ababa on 19 January 1946 and helped to organise the Cadet School. These units were

6 Notable is that Swedish (Forsgren, SAFO No. 20) and Ethiopian sources differ over the base of the IEAF's Flying School: the Swedish state Harar Meda near Addis Ababa, while the Ethiopians insist this was actually stationed at Dire Dawa, a town of around 70,000 where a major army base and supply depot were subsequently established too.

7 Author's notes, based on various publications specialised in aeronautical affairs. Von Rosen later became prominent for his actions in Biafra in 1968. He returned to Ethiopia in 1977, this time pioneering the use of MFI-15 aircraft for air drops of food supplies during a catastrophic famine. He was killed in an attack by Somali guerrillas.

Pre-delivery photograph of seven SAAB 91 Safirs, which served as primary trainers until 1973. No less than 48 were purchased by Ethiopia during the 1950s. (EtAF)

to be equipped primarily with Swedish aircraft, including Svenska Aeroplan Aktiebolaget (SAAB) 91 Safir training aircraft, SAAB B.17 fighter-bombers and FFVS J.22 fighters.[8] Furthermore, he recognised that the existing Lidetta airport outside the capital was insufficient for planned future operations. Therefore, a decision was taken to construct a new air base at the race course in Bishoftu, situated near the former Italian experimental agricultural station and a railway station on the Addis Ababa to Djibouti railway. Working together, Ethiopian cadets and their Swedish instructors constructed the new airfield on time for the arrival of the first six SAAB 91 Safir training aircraft on 24 December 1946. Nearly a year later, on 10 November 1947, Swedish pilots also delivered the first 16 SAAB B.17 fighter-bombers to Bishoftu, and subsequently the Flying School moved there. The new air base was officially inaugurated on 20 May 1947, although this ceremony was formally repeated for Emperor Selassie on 13 November of the same year, when the flight school was officially inaugurated. This event also marked the graduation ceremony of the first generation of Ethiopian pilots, technicians and radio operators, as well as the official introduction to the service of the recently delivered aircraft obtained from Sweden, and the naming of the air base as 'Harar Meda' ('Harar Field'). Curiously, Harar Meda soon became colloquially known as 'Debre Zeit', after a nearby town constructed around the time to house the families of IEAF personnel.[9]

Except for contracting Swedes to help establish the air force during the 1950s and 1960s, the Ethiopians also recruited British, Norwegians, French, Indians and Israelis to help organise, train and advise a small navy. Israeli advisers conducted special infantry training and provided advisers for the Frontier Guard (responsible for monitoring the border to Somaliland) and the Commando Police. West Germany provided equipment for police field units, while India helped with the training of the Imperial Bodyguard and the establishment of the faculty of the Military Academy at Harar.[10] Ethiopian officers attended military schools in the USA, UK and Yugoslavia, while a volunteer battalion of the Ethiopian Army, consisting of the Imperial Bodyguards, and better known as the 'Kagnew Battalion', was deployed in Korea together with UN forces, where they fought with distinction.

8 Forsgren, SAFO No. 20, p. 23.

9 *Local History in Ethiopia*, released by the Nordic Africa Institute website, accessed June 2009; Forsgren cites that as of 13 Nov. 1947, the IEAF consisted of sixteen B. 17 As, five Safirs, four Tiger Moths, two Cessna UC-78s, and one Avro XIX.

One of the Fairey Firefly FR.Mk 1s operated by the attack squadron of the IEAF in the 1950s and early 1960s. (EtAF)

Safirs and Fireflies

Relations with Sweden remained crucial for the development of the IEAF in the 1950s. Not only did the air force remain under command of a Swedish officer, but it continued purchasing Swedish aircraft, partially because of a high attrition rate. This resulted in Ethiopia eventually acquiring a total of 48 Safirs (between them 9 Lycoming-powered SAAB 91Bs) and 46 SAAB B.17A light bombers (a few of which were locally modified to the Sk.17A training variant, and one of which was assembled in Ethiopia from spare parts).[11]

10 Dupuy et al., p. 207.

11 Forsgren, SAFO No. 20, p. 24.

Nevertheless, hoping for a more potent combat aircraft, in 1950 the IEAF also contacted Fairey Aviation in the UK, expressing an interest in obtaining up to 35 Firefly fighter-bombers. Following some negotiations, a contract was signed for nine of these, in early 1951, including eight Firefly FR.Mk 1s and a single Firefly T.Mk. 2 training aircraft. These eventually entered service with the then only operational IEAF unit, the Harar Meda based 'Attack Squadron', starting in early 1952. Two years later Ethiopia acquired 14 additional Fireflies, which had formerly served with the Royal Canadian Navy, and entered service with an attack squadron based at Asmara.[12] Both units flying Fireflies frequently suffered from lack of spares and maintenance related problems, lack of qualified technicians and pilots (there were never more than 12 pilots to fly all the Fireflies), and the IEAF never operated more than 16 such aircraft at the same time. In fact, most of the Fireflies delivered to Ethiopia never officially entered service, but were used as sources of spares immediately after their delivery.

The idea of establishing a tactical attack wing of three squadrons equipped with this type and B.17As thus never materialised, no matter how intensively the pilots trained. Although the number of Swedes serving with the IEAF had increased to 143 by 1956, and von Rosen, who was actually a civilian pilot, was meanwhile replaced by a professional Swedish military officer, the Flying School remained a relatively small institution concentrated on producing pilots of high quality. It continued deploying SAAB Safirs for primary instruction, and Cessna AT-17 Bobcat twin-engined trainers for advanced tuition. These training phases lasted a total of one year at Debre Zeit, after which the future IEAF pilots were provided with a second or even third year of operational conversion. While all IEAF pilots were fully qualified to operate from small and rough airstrips at high altitudes and ambient temperatures, the annual output of some 20 new pilots established by their Swedish instructors through most of the 1950s was never increased even though three British instructors were also on the strength and later some locals qualified for this duty. The only exception to this rule was when the Flying School took over the duty of providing basic training for Ethiopian Airline pilots.[13]

While the development of combat assets thus stagnated, the purchase of several de Havilland Doves from the UK, and two Stinson L-5 Sentinels from the USA enabled the establishment of a transport and communications squadron in 1948. This unit was reinforced through the addition of two Douglas C-47s in 1955, which proved capable of operating from nearly all the strips in the outlying country. Furthermore, the IEAF constructed additional air bases in Jijiga, Dire Dawa and Asmara.

US Military Aid

Following corresponding negotiations, the Ethiopian and US governments signed an agreement about military cooperation and a Military Aid and Assistance Group (MAAG) was posted to Addis Ababa in 1953. The work of the MAAG was primarily related to the development of the Ethiopian Army and Navy in exchange for the Americans receiving rights to re-open the radio-relaying and receiving station in Kagnew, outside Asmara. Originally constructed in 1942, this facility was crucially important for communication between Europe and the Far East during WWII, and the Americans found it perfectly positioned during the military build-up prompted by the outbreak and spread of the Cold War into Southern Asia and Africa. Correspondingly, approximately 300 US service personnel and soldiers were stationed in Ethiopia. In April 1957, members of the MAAG concluded a comprehensive study of Ethiopian military capabilities, requirements and possible threats. The head of the team, Col Robert L Ramzi, concluded that the IEAF consisted of three fighter-bomber squadrons equipped with obsolete and slow B.17s and Fireflies, and one transport squadron equipped with a miscellany of aircraft. He recommended the air force be reorganised to include one wing with 36 fighter planes and one transport squadron with eight aircraft. Following additional negotiations, this study was converted into an 'implementation plan' that envisaged for the IEAF to be completely expanded by 1959, as follows:

1) 25 pilots were to undergo conversion courses for jet fighters in the USA;
2) the fighter-bomber wing was to be equipped with three squadrons of jet fighter-bombers;
3) the transport squadron was to be completely re-equipped with C-47 transports;
4) US military assistance would primarily concern training-related issues, especially so in the case of the IEAF.[14]

This plan was accepted by Emperor Selassie and his military commanders and the Americans worked quite fast. By the time Ethiopia received the first out of an eventual 13 C-47s in 1956, the

Debre Zeit AB as seen around 1960, with several hangars still under construction. Visible are (from left to right) two of recently delivered T-33s, five Safirs, and four C-47s. (EtAF via S.N.)

Pilots of the 21st Transport Squadron in front of a C-47 operated by that unit. (EtAF via S.N.)

12 Hellström, Imperial Fireflies; reports about acquisition of twelve ex-Dutch FR.Mk. 1s during the same year were apparently based on Ethiopians expressing interest in these aircraft. However, they were never purchased, see Forsgreen SAFO No. 20, p. 24.

13 Forsgreen SAFO No. 20, p. 24 & Green et al., p. 89.

14 Mekonnen, pp. 99–100.

A rare photograph of two Canadian-built RT-33As, which are known to have entered service with the 33rd Squadron in the early 1960s. They were left in highly polished 'aluminium' overall. Except for roundels (applied in six positions) and the usual serials in black, they only wore the usual set of maintenance and warning stencils. (EtAF)

Some of first IEAF technicians re-qualified to maintain F-86Fs, posing in front of a bomb-armed Sabre, in the early 1960s. (via S.N.)

MAAG had already trained enough pilots and technicians to operate these and began re-orienting to convert IEAF pilots and technicians to jet fighters.[15] Correspondingly, a large team of US Air Force (USAF) instructors arrived at Debre Zeit Air Base (AB) in 1957, together with the first three out of an eventual 14 Lockheed T-33A T-Bird jet trainers, and four additional C-47s.[16] Once in place, the Americans trained Ethiopian pilots, and also a few Swedes, on both types. On 19 February 1958, an additional group of Ethiopian pilots departed for Randolf AFB, in the USA, where they were to undergo conversion training on North American T-28As, and then T-33As. The second group, including six experienced pilots that were to be trained as instructors and two groups of cadets fresh from primary military and flying training, were sent for advanced courses to the USA on 28 February 1959, while the first group returned to Ethiopia by 20 December 1960.

Although these and most subsequent courses for Ethiopian pilots and ground personnel were organised at airfields in southern USA, in federal states characterised by racism and segregation, thus making matters neither easy nor pleasant for Ethiopians, the IEAF personnel completed their training and passed all exams with very high marks. Not a single student failed and the group was back in Ethiopia by 20 December 1960. Unsurprisingly, US instructors that used to train IEAF pilots on T-28A/Bs and T-33As in 1958, recalled very favourably about them.[17] At the same time however, there were several failures on the part of Swedish students trained by Americans together with Ethiopians on T-33s, and this factor reportedly created quite some tensions within the IEAF.[18] More tensions were to emerge quite soon as US aid continued to accelerate the development of the air force. Late in 1959, the attack squadron at Asmara was deactivated. Most of its Fireflies had already been grounded since the previous year, even though some were briefly returned to service in 1960, when there were the first signs of tensions with the newly independent Somalia and the IEAF was ordered to 'show the flag' along the border region.[19] Instead, the newly established 1st Squadron converted to T-33s under supervision of US instructors and moved to Debre Zeit.

Meanwhile, the group of six Ethiopians sent to the USA in 1959 had successfully completed their conversion on T-33As and the instructor-pilot-course at Luke AFB. Conversion training then began on North American F-86F Sabre fighter-jets, 10 of which were supplied to Ethiopia between September 1960 and June 1967. With the help of US instructors, six IEAF pilots were trained on the type at Debre Zeit AB within only four months, and in early 1961 the 1st Fighter Intercept Squadron was activated at that base. In 1962, eight T-28A/Bs entered service with the 32nd Squadron, responsible for intermediate training (this unit was reinforced by 14 additional examples delivered in 1964), while T-33As were handed over to the 33rd Squadron, responsible for jet training. Furthermore, through the first half of the 1960s, the USA provided enough C-47s, followed by two Douglas C-54 Skymaster transports for the IEAF to organise them into the 21st Transport Squadron.

The reason for the flawless, even though relatively slow, conversion process of IEAF personnel from piston engined to jet fighters and US-made transports in the USA and at Debre Zeit AB, is quite obvious. As in previous times, because of the limited financial resources of their country, the Ethiopians continued emphasising quality over quantity. Furthermore, and in agreement with the Americans, they intended to build-up an embryonic force which was to serve as a catalyst for future growth. Therefore, the Ethiopian military was accepting only the best and fittest candidates. However, it was for this reason that the subsequent development of the IEAF, and the rest of the Ethiopian military, proceeded at a rather slow pace. For most of the early 1960s, the emphasis on quality over quantity resulted in a decrease of output of new officers and pilots to a level where only between five and ten students were graduating from the Cadet College of the IEAF annually. Therefore, although a large group of Ethiopian pilots returned from two years of education in the USA, and 28 additional ex-USAF F-86Fs were supplied by 1970, it would take the IEAF many years to establish additional flying units.

Coup Attempt and Deployment to Congo

On 14 December 1960, the Swedish-trained Imperial Body Guard revolted against the Emperor while he was on a visit to South America, claiming their intention to bring democracy to the country and end widespread corruption. As the fighting erupted between the guard and the army in Addis Ababa, the last Swedish commander of the IEAF, Brigadier General (Brig Gen) Knut Lindahl, in agreement with members of the MAAG, ordered the air force into action. The B.17s and F-86s flew a number of attacks on rebel positions in the capital

15 The first Ethiopian pilot to fly a jet fighter was one of the pilots that did an exchange tour with the Swedish Air Force, in 1956. Sadly, he was killed when the de Havilland Vampire of the Halmstad-based F14 squadron crashed, during the same year, see Forsgren, SAFO No. 20, p. 24.

16 The first three T-33s were followed by two in 1962, five in 1964 and four in 1966.

17 Tom Long, son of one of the former USAF instructor pilots who trained Ethiopian pilots on T-33As and later F-86Fs, e-mail interview, Oct. 2003.

18 In an 'off the record' part of an interview, one of the participating Ethiopian pilots concluded that the Swedes were apparently not sending their 'best and brightest' pilots to serve as instructors in Ethiopia.

19 Hellström, Imperial Fireflies.

Ethiopian and Irish officers in front of one of four IEAF F-86Fs deployed to Congo, as seen at Leopoldville International on 3 October 1961. (UNmultimedia)

during the following two days, apparently all flown by Ethiopian pilots, although Lindahl probably made one reconnaissance flight over Addis Ababa. The coup attempt eventually collapsed. However, together with already existing tensions between Ethiopians and their Swedish instructors, it prompted Emperor Selassie to order the withdrawal of the Swedish mission to the IEAF. The Swedes were not interested in continuing their involvement either and, on 8 May 1961, they informed the IEAF that no further personnel would be made available. Therefore, Brig Gen Assefa Ayene (formerly an IEA-cadet from the 1930s, and then the Deputy Commander of the IEAF Training Centre), became the first Ethiopian C-in-C Imperial Ethiopian Air Force, replacing Lindahl before the end of the same year.[20]

Meanwhile, although often called to show the flag to Somalis during some of numerous border clashes of the following years, the IEAF saw its next action between September 1961 and October 1962, when it deployed a flight of four F-86Fs with UN-peacekeepers in Congo, and in support of some 3,000 Imperial Bodyguard personnel (about 10% of the then entire Ethiopian military strength) that served there as well. Drawn from pilots assigned to the 1st Fighter-Interceptor Squadron, the four F-86s took off from Debre Zeit AB on 25 September 1961 and deployed to Leopoldville (now Kinshasa) via Entebbe in Uganda and Stanleyville (now Kisingani). A support team with spares and tools was brought to the Congo by Lockheed C-130 Hercules transports of the USAF.

This journey was quite challenging for young Ethiopian pilots, not only because of frequent and sudden changes of weather, but particularly because of the short runway in Stanleyville, surrounded by trees. Nevertheless, they had already proved their worth days before their arrival, and this without firing a single shot. Foreign mercenary pilots, flying for the 'air force' of the secessionist Katangan government in southern Congo, began demanding higher payments, citing the appearance of Ethiopian fighter jets as the reason.[21]

After operating out of Leopoldville for two weeks and in conjunction with contingents from the Swedish and Indian Air Forces, the IEAF detachments moved to the huge Kamina AB in southern Congo, constructed by the Belgians in the 1950s with the intention of evacuating their government in the case of a nuclear war in Europe. Together with SAAB J-29 Tunnans and English Electric Canberra bombers, they became involved in the war against Katangan separatists which, amongst others, operated two armed Fouga CM.170 Magister jet trainers of French origin and a number of T-6 Texans. While involved in combat missions, when they were often fired upon, Ethiopian Sabres suffered no losses, but one fighter crashed on 14 October 1962 in unknown circumstances. The other three F-86Fs returned to Debre Zeit eleven days later, their withdrawal being kept so secret by the UN HQ that even the co-located Swedish unit in Kamina did not know about it until the Ethiopians took off for their return flight, on 25 October 1961.

Two years later, another detachment of IEAF Sabres was deployed to Congo again, but this time they saw no combat operations and were withdrawn within only three months.

Origins of the Somali Air Force

While administrating the UN Trust Territory of Somalia in the 1950s, the Italians established the Corpo Aeronautico della Somalia (CAdS), an embryonic air arm, the task of which was to support the authorities and aid establishment of the future Somali Air Force. This organisation was equipped with a miscellany of aircraft, primarily from Italian stock, and including two Douglas C-47 Skytrains and eight C-53 Skytroopers, two Beech C-45 Expeditors, two North American T-6H Texans, two Stinson L-5 Sentinels, and six North American P-51 Mustang piston-engined fighters.[22] Although manning these aircraft with their crews, the Italians trained 32 officers of the Scout Regiment (a unit that subsequently became the backbone of the Somali National Defence Force, SNDF), as pilots and technicians. This started in 1954, mostly in Italy, where some qualified to fly C-47s and Savoia Marchetti S.82 transports, while others qualified as maintenance technicians at the Industrial School in Napoli. At the same time, the CAdS adapted the future Somali national insignia, including a white star on blue background, originally used by the southern part of the country.[23]

Although still under Italian sponsorship, and adapting the same national markings as previously worn by CAdS aircraft, the fledgling air force was re-organised as the Somali Aeronautical Corps (SAC) in 1958, by which time it operated the Flying School in Hargheisa, as well as one fighter and one transport squadron, both based in Mogadishu. Although all the surviving Mustangs were returned to Italy before Somalia was released into independence, this organisational structure was retained until well into the mid-1960s.

By the time Somalia gained independence, the SNDF and the SAC consisted of only around 5,000 officers, non-commissioned officers (NCOs) and other ranks. Indeed, the backbone of the SNDF – then commanded by Major General (Maj Gen) Daud Abdullah Hersi – was still the sole scout regiment, originally established by the British in the 1940s. The SAC, then under the command of Col Ali Matan Hashi, consisted of a core of reasonably experienced and Italian trained officers and other ranks, but lacked operational aircraft. As of 30 June 1960, the Italians had handed over a total of only three C-47s, two Beech C-45s, one T-6, one L-5,

20 Forsgren, SAFO No. 20, p. 24.
21 Mekonnen, p. 117 & *Tom Tom* (UN weekly newspaper in the Congo), various volumes.
22 Breffort et al., *The North American P-51 Mustang: from 1940 to 1980* & Brent, p. 143.
23 Wheeler, p. 85. Note that other authors identified Somali T-6s as 'AT-6 armed trainers'.

The first group of Somali cadets selected for training to become pilots and technicians in Italy in 1954. (Sheikh Hassan Collection)

Tahsin Zaki, an Egyptian instructor pilot, with two Somali cadets. Visible on the fin of Gomhouria trainer behind them is the original fin flash as worn by SAC aircraft in the 1960. (Tahsin Zaki via Dr David Nicolle)

Although the Italians trained the original cadre of Somali personnel, additional cadets were subsequently trained either in Egypt or Iraq. This was one of two Gomhouria basic trainers donated to the CCS. (Tahsin Zaki Collection)

and one Piper in operational condition.[24] Nevertheless, although the first flight by an all-Somali aircrew was recorded only in early 1960 (on the morning of 15 February), by June there were enough pilots and technicians to man and service all of these aircraft.[25]

By the time the SAC had been officially renamed as the Ciidanka Cirka Soomaaliyed (CCS, the Air Force of Somalia) in December 1960, it was reinforced through the addition of two Egyptian made Gomhouria trainers, and had already several groups of cadets in training to become pilots and technicians, posted to Egypt, Iraq, and Italy. Except for Gomhourias, the CCS was reinforced through eight Piaggio P.148 basic trainers donated by Italy in 1962, which enabled the re-equipping of the Flying School in Mogadishu. However, it was the ties between Somalia, Egypt and Iraq that resulted in the most important early reinforcements. During the following years, the CCS gained six Somali pilots, including Mohammad Barre 'Haytan', who completed their training at the Iraqi Air Force Academy at Wahda AB, near Basrah, becoming qualified fighter pilots in November 1961, and several pilots graduated at the United Arab Republic Air Force Academy at Bilbeis Air Base, in Egypt. Baghdad donated three de Havilland Vampire T.Mk. 11 jet trainers and Cairo two additional Gomhourias in 1963.

Soviet Ties

Subsequently, expansion of the SNDF and the CCS became an urgent requirement for the government in Mogadishu, which came under severe pressure from different parties, but especially that of the 'Somali Youth League', a social and political movement established in 1942 to provide a union of all areas predominantly populated by the Somalis. Frankly speaking, this meant that during the following years, the foreign policy of Somalia largely consisted of attempts to achieve more than only 'symbolic' unity between the nation and various groups of ethnic Somalis living abroad, i.e. considered as 'living under foreign control' by the majority of politicians in Mogadishu. To symbolise this 'historic' mission, the Somali flag was emblazoned with five stars, two representing the former Italian and British colonies now constituting the Somali republic and each of the others representing one part of the greater Somali nation.

Correspondingly, the government of President Daar not only refused to acknowledge the validity of the 1954 Anglo-Ethiopian treaty recognising Ethiopia's claim over Ogaden, or any of the treaties made with Somali clans that put themselves under foreign protection, but the relevance of all the possible treaties defining borders to neighbouring countries. Officially, Mogadishu concluded that Somalis were never consulted on the terms of these treaties and as far as they were concerned, they had not been informed of their existence.[26] Determining that the crucial 'Somali national interest' was 'recovery of the lost territories', and that this was possible only by force, Daar and his government decided to develop a much more potent 20,000-man strong army, and requested support from the USA and Italy. Arguing that such a military would be insufficient to maintain the internal security of Somalia, Washington offered, in cooperation with Italy, a joint military aid project limited in size and scope, which Mogadishu rejected as 'too small and too restrictive'. Instead, Somalis began looking elsewhere for military assistance.[27]

It was under these conditions that Moscow, which was seeking to counter Western influence in Africa throughout the 1950s, established ties to Mogadishu and granted an 'unconditional' loan of US$32 million with a fixed 20-year repayment schedule, aimed primarily at expansion and modernisation of the Somali military.

24 Ahmad Shaikh Hassan (former CO Flying School in Hargheisa), interview, Dec. 2008.

25 See also Brent, p. 143, for details. Note that Brent counted one C-53 and only two C-47s as available on independence, while Somali sources indicate no presence of any C-53s, and availability of three C-47s. Probably, the Somalis made no distinction between the C-47 and C-53.

26 Nkaisserry, p. 9.

27 Dupuy et al., p. 256.

Mualim Abdullah 'Fojar' in the cockpit of one of CCS' MiG-15UTIs in the late 1960s. (Sheikh Ahmad Collection)

Group of CCS officers and pilots serving with the Flying School at Hargheisa in the late 1960s. Standing fourth from the right is Asli Hassan Abade, the first Somali female pilot. (Sheikh Hassan Collection)

Correspondingly, while the SNDF began ordering enough main battle tanks (MBTs), armoured personnel carriers (APCs), artillery and other armament to expand to 14,000 personnel, the SAC ordered ten Yakovlev Yak-11 basic trainers, eight MiG-15UTI jet trainers, 40 MiG-17 fighters, and at least three An-2s from the USSR in 1963.[28]

Acknowledging that the SNDF and the SAC would require several years to train sufficient personnel and receive all of this equipment, Moscow then increased its loan to Somalia to US$65 million, demanding and receiving base rights for port facilities in Berbera, barely 250km from the strategic Bab el-Mandeb Straits and the (then British-held) port of Aden. Mogadishu accepted this condition because it enabled orders for additional aircraft, including three Antonov An-24 transports and four Mil Mi-4 helicopters, one air defence brigade equipped with SA-2 SAMs and tanks, but also because the Soviets signalled their preparedness to reconstruct all the existing, or build several new, airfields, including Mogadishu, Hargheisa, Baidoa, and Kismayu. Furthermore, in order to accelerate training and conversion of Somali personnel to Soviet equipment, around 300 Soviet military advisers were deployed to Somalia, while several thousand Somalis, including around 500 pilots, officers and technicians, underwent training in the Soviet Union. In this fashion, and although not in the least 'communist' by its political orientation, the government in Mogadishu established particularly close ties to the USSR.

As mentioned above, at least some of the early Somali fighter pilots sent to the Soviet Union for conversion courses on MiG-17s, starting with late 1964, were already in possession of some experience on jet aircraft. Once in the USSR, they first underwent a lengthy course in Russian language and then re-qualified on Aero L-29 Delfins and MiG-15UTIs, before continuing their training on MiG-17s at Borisoblebsk Air Base. More experienced pilots and technical officers intended to serve in top positions, received additional education at the Air Defence School in Armavir which included more theoretical studies, but also some live firing exercises with night and all-weather flying training that prepared them for deploying MiG-17s in combat. Later during the 1960s, further Somali pilots also attended courses for fighter-bombers in Eysk (Taganrog). Regardless of their previous experience, on their return to Somalia, Somali pilots honed their skills in regular exercises. Throughout this time, all the novice cadets received basic flying courses in Egypt.

First Ethiopian-Somali Clashes

During the mid-1960s it became obvious that Daar's efforts to expand the military and raise its combat effectiveness to a level where it would be able to fight wars for 'liberation' of ethnic Somalis living outside Somalia's borders, could never proceed at the pace demanded by various hardliners. Because most of the Soviet loans were spent for purchasing arms and training military personnel, the economy remained weak. In return, lack of interest and investment from abroad resulted in the nation's inability to purchase more arms within a shorter period of time.

All the possible weaknesses of the Somali military became more than obvious during the first five years of Somalia's independence. Tensions were rampant along the border with Ethiopia, as well as inside Ogaden, and minor clashes between Ethiopian police and armed parties of Somali nomads, further complicated by traditional movements of Somali nomadic herdsmen across the poorly defined demarcation lines, were frequent. Hostilities grew steadily in scope, eventually involving small scale actions between the armed forces of both countries. However, the weakness of the Somali military became especially obvious during the war that erupted in February 1964 along the entire length of the border. Details about this short conflict are scarce but it transpired that the IEAF's T-33As and RT-33As not only flew reconnaissance sorties 'showing the flag', but the F-86Fs also heavily bombed Somali positions near the border town of Togochale in Somalia. The Somalis reportedly deployed their Vampires to fly combat air patrols (CAPs) along the border, while their C-47s transported two companies of ground troops into the crisis zone. However, generally the Ethiopians encountered very little resistance in the air and on the ground. In fact, Ethiopian Sabre-pilots recall flying CAPs as deep inside the Somali airspace as Hargheisa, without encountering any opposition in the air. Open hostilities were brought to an end in April 1964, through the mediation of Sudan, which acted under the auspices of the Organisation of African Unity (OAU). A new series of border clashes followed in October of the same year, and again proved that the Ethiopians were vastly superior to the inexperienced Somalis.

Eventually, President Daar had to acknowledge that Soviet aid was insufficient to match Ethiopian military power for the next few years. His country was actually largely dependable on donations, of

28 Various sources differ strongly over the number of MiG-17s ordered (though usually claimed as 'delivered') by Somalia in 1963 and 1964, from the twelve usually mentioned, to the most frequently mentioned figures of between 30 and 40, and up to 54 (combined number of MiG-15UTIs and MiG-17s) according to *Aerospace Encyclopaedia of World Air Forces*, p. 146. The figures provided here are mostly based on interviews with former CCS pilots, who added that the 40 MiG-17s delivered in the given period were eventually distributed between no less than four fighter squadrons.

Technicians of the 1st Fighter-Interceptor Squadron in front of one of the F-86Fs they were servicing during the short war with Somalia in 1964. Kneeling (f.l.t.r.): Metekia Dibaba, Tessema Mamo, Teklu Cerinet, Tekola Astatke, Abate and Alemayehu Getachev. Standing (f.l.t.r.): unknown, unknown, Assefa Birbo, Mitiku Wolde, Endale Derseh, Beyene Berhe, Zeleke Haile Mariam, and Alemayehu Asfaw. (S.N. Collection)

Capt Bekele Gululat, Capt Tassew Gebre Medhin and Lt Girma Bekele with one of the IEAF's B.17s at Kebridehar airfield, in Ogaden, during the war with Somalia in February 1964. (EtAF via S.N.)

One of the CCS pilots trained in Egypt and Iraq with an ex-Iraqi Vampire T.Mk 11 seen while on alert on 15 October 1964. (Sheikh Hassan Collection)

which only a few ever materialised. With his policy failing, Daar was defeated during elections in 1967, and Prime Minister Abdi Rashid Ali Shirmake took over as a new president.

Freedom Fighters

With Ogaden becoming the main bone of contention between Ethiopia and Somalia, but also because of the spread of the pan-Arabic wave through the Middle East and the growing arsenal

Major Yohannes Woldemariam with the first F-5A delivered to Ethiopia in 1966. (EtAF via S.N.)

of communist arms in Somalia, Addis Ababa and Washington concluded that there was a serious threat to Ethiopian territorial integrity. This in turn stimulated a significant military build-up in Ethiopia during the second half of the 1960s.

The Ethiopians monitored developments in the neighbouring country with increasing concern. The arrival of MiG-15s and MiG-17s was something the F-86-equipped IEAF could tackle, but Somali acquisition of a large number of T-34 tanks created some concerns, especially because the US administrations of Presidents Dwight Eisenhower and John F Kennedy refused to match these through deliveries of comparable equipment to Ethiopia. The situation began to change once Lyndon B Johnson became the President. Following negotiations with a number of his officials, Emperor Selassie began requesting additional armament and military aid from Johnson; the then Secretary of Defence, Robert McNamara, eventually came up with idea to equip the IEAF with a squadron-worth of Northrop F-5A Freedom Fighters, which were expected to reinforce Ethiopian defence capabilities without further escalating an already tense situation.

The original IEAF team that converted to F-5As consisted of highly experienced senior pilots and instructors, led by Maj Yohannes Woldemariam.[29] Their training on F-5s was undertaken under the auspices of the 4441st Tactical Fighter Wing (TFW) at Williams Air Force Base (AFB) in Arizona, in 1965. It began with an English language course (where necessary), followed by some 40 hours of flight training, 115 hours of classroom instruction and 50 hours of 'briefing and critiques'. Pilots then flew about 15 hours of instruction in two-seater F-5Bs and 25 hours on F-5As, the curriculum covering formation flying, instrument flying, air-to-air and air-to-ground gunnery, missile attack profiles and air combat manoeuvring. Together with Tigneh Woldegiorgis, while his junior colleagues were still undergoing their conversion courses, Woldemariam further underwent combat crew and instructor training on the F-5A. Except for Ethiopian pilots, USAF instructors at Williams AFB also trained IEAF technicians, passing on their expertise in the whole range of aircraft support and maintenance skills during courses that varied from 85 to 250 hours, depending on the complexity of the subject. All the training was ably assisted

29 The memorial website for General Yohannes Woldemariam (in Amharic) states that he could be described as a 'prototype' of an Ethiopian fighter pilot that, 'pursued rigorous professional education in the USA and Great Britain from 1960, and attended the USAF Air University'. Note that, due to uncertainties over exact military ranks of mentioned pilots at the given time, the author decided to leave these out.

by simulators and advanced training aids. [30]

Therefore, while the Somali MiG-17-pilots never received more than basic conversion courses in which they only learned how to take-off and land their aircraft, the Ethiopians received all the instruction necessary to deploy their F-5s in combat and to convert additional pilots and ground personnel at home.

Advanced Training

Approximately three months before the arrival of first F-5s in Ethiopia, the MAAG and the IEAF, assisted by a USAF Mobile Training Team (MTT) began working on establishing the infrastructure necessary to support F-5 operations at Debre Zeit, ensuring that there was a minimum of delay in getting the new aircraft operational once they arrived in Ethiopia.[31] Once he returned to Debre Zeit, Woldemariam became the first CO of the appropriately designated 5th Fighter-Interceptor Squadron, also known as 'Tiger Squadron'.[32] Together with Woldegiorgis, he then trained additional Ethiopian pilots, including Berhanu Wubneh and Teshale Zewdie, to become instructors for F-5s, with MAAG advisers and several Northrop-representatives serving as 'quality controllers'.[33] In turn, during the following months and years, Teshale, with aid from the MAAG team, provided extensive conversion and courses in basic fighter manoeuvring, air combat manoeuvring, tactics and air-to-ground gunnery. The first group of Ethiopian F-5 pilots converted at Debre Zeit included Ashenafi Gebre Tsadik, Berhane Kebede, Girma Workagexehu, Haile Michael Birru, Lagesse Teferra, Naizghi, Yohannes Aregawi and Tigneh Woldegiorgis; the second included Berhanu Wubneh, while junior pilots – including Afework Kidanu – trained in the USA around the same time.[34]

The first of F-5As and F-5Bs arrived in Ethiopia in 1966. Although equipped to the same standard as F-5s delivered to other foreign customers at that time, they were never wired for AIM-9B Sidewinder air-to-air missiles. This was prevented by the US State Department, which embargoed delivery of such weapons to any African nation perfectly in accordance with McNamara's intentions in regards to Ethiopia.[35] For this reason, the two Colt 20mm cannons remained their only weapon for air-to-air combat, while 75mm (2.75in) unguided rockets, as well as M-117 and Mk.82 GP-bombs, napalm bombs and cluster bomb units (CBUs), were supplied for air-to-ground missions.

The operational ready rate of IEAF F-5s quickly reached 85%, which was slightly above average in comparison to other contemporary foreign customers, such as Iran and South Korea. Unsurprisingly, the training of pilots and ground crews proceeded at a high pace and proved very successful, although one aircraft and its pilot, Yohannes Aregawi, was lost in the course of an air combat training exercise against a US instructor, undertaken at a very low level. One of the F-5Bs was also written off at an unknown date, some eleven to twelve months later. After the minimum altitude for such exercises was set at 4,520m (15,000ft), no other Ethiopian F-5s was lost in any kind of accidents until 1977.

In 1968, two IEAF F-5-pilots were sent to the USA to undergo

Emperor Selassie inspecting one of the IEAF F-5As upon their delivery to Debre Zeit AB in 1966. (EtAF)

Group of pilots of the 5th Squadron in front of one of their mounts. Standing, left to right: Fikru Maru, Ashenafi Gebre Tsadik, Belay Teklehaimnot, Techane Mesfin, Berhanu Wubneh, and Techale Zewdie. Kneeling, left to right: Girma Workagexehu, Estifanos Mekonnen, Addis Tedla and Ambachew Wube. (via S.N.)

30 Scutts, pp. 29–30.

31 The first F-5As arrived in Ethiopia in 1966, flown directly from McClellan AFB, in California. That was probably the longest delivery flight ever flown by F-5s.

32 Designation of the 5th Fighter-Interceptor Squadron is provided here as cited by one of its former USAF instructors, who granted an interview on condition of anonymity. Some Ethiopian sources cited the actual designation of this unit as the '5th Tactical Fighter Squadron'.

33 Note that Ethiopians are identified by their first names. For example, Berhanu Wubneh was usually referred to as 'Berhanu'; Teshale Zewdie as 'Teshale' or 'Lieutenant Teshale'.

34 Ashenafi Gebre Tsadik, interview, 2001. Specifically, Ashenafi recalled that an average IEAF pilot of that period first had to undergo around 100 hours on Safirs and 100 on T-28s during Phase I of his training (basic flying training; Ashenafi's T-28 instructor was Berhanu Wubneh); around 100 hours on T-33s during Phase II (jet conversion training); around 30 hours of combat training, including air-to-air and air-to-ground gunnery; six hours conversion course on F-5Bs and then many more on F-5As. Quality standards were excessively high, drop-out rates regularly surpassing 90%: very often, only four out of 48–50 candidates managed to qualify as jet fighter pilots. Furthermore, once qualified, each pilot was expected to fly a minimum of sixteen hours a month (instructor pilots clocked as many as 30 a month), and had to pass gunnery qualification every three months. Additionally, IEAF pilots participated in frequent exercises with the army, resulting in them regularly flying 200 hours a year.

35 *Action Memorandum*, by David D Newsom, US Department of State, 5 Jan. 1974, provided in response to FOIA inquiry.

The first of four Canberra B.Mk 52s seen during a pre-delivery test-flight over the UK in 1968. (BAe Heritage)

the survival training course, undertaken in Washington State. There they received basic training, as well as training in escape and evasion. The Ethiopians were excluded only from the 'resistance' portion of this course, which remains classified for all foreign trainees even today. In turn, once they were back in Ethiopia, the pilots in question established a survival school and qualified as instructors.

Bomber Dilemma

The arrival of F-5s at Debre Zeit provoked a furious reaction from the government in Mogadishu, which described their appearance as a grave threat to the security of Somalia and the entire Horn of Africa. The Somali government delivered a 14-page letter of protest to Washington, but this action backfired. The Americans reacted through their Ambassador to Somalia; he approached his Soviet counterpart and entered a 'gentleman's agreement' not to supply any more advanced weapons to the region for the time being.[36] For as of yet unknown reasons, Moscow swiftly forgot about this agreement and delivered four Ilyushin Il-28 bombers, one An-26B transport and five Mil Mi-8 helicopters to Somalia, between 1967 and 1972.

Mainly flown by former MiG-17-pilots, Il-28s provided the CCS with a capability to hit targets anywhere in Ethiopia, unmatched by anything in service with the IEAF. Reports about Somali requests for delivery of such bombers prompted the Ethiopians to completely re-think their preoccupation with 'quality over quantity' which caused the slow output of new pilots and their training at home. Because it lacked aircraft capable of reaching all parts of Somalia, the IEAF demanded the acquisition of a similar capability. With the USA turning down all the corresponding requests, primarily because several years were to lapse before the Soviets indeed delivered the first two Il-28s to Somalia, the Ethiopians turned to Great Britain and ordered four English Electric Canberra B.Mk 2 bombers, in 1968. Washington strongly opposed this idea and the Americans tried their best to dissuade the British from supplying such aircraft, causing quite some controversy in Addis Ababa. Nevertheless, Emperor Selassie pushed the deal through and after being overhauled and adapted for counterinsurgency (COIN) role, the Canberras were delivered to Debre Zeit under the designation B.Mk 52, forming the unit usually known only as 'Bomber Squadron', but officially designated the 44th Bomber Squadron. Most of the Ethiopian Canberra pilots, including Bezuwork, Assefa Mekbib, and Afeworki Mekonnen used to fly T-33s at earlier times, but one, Mesfin Haile, was a former F-86-pilot.

First Blood in Eritrea

Except for the increasing threat emerging from Somalia, another reason for the military build-up in Ehtiopia during the mid-1960s was the slowly spreading insurgency in Eritrea. Colonised by the Italians in 1882 and occupied by the British since 1941, Eritrea came under a UN mandate in 1950 because the Soviet Union stunned Western Allies by requesting trusteeship over that area during the Potsdam Conference in 1945. After the British withdrawal, Eritrea came under full Ethiopian control, and in 1955 Emperor Selassie federated the area to become one of nine provinces making the Ethiopian union. Not all Eritreans found this decision acceptable. Primarily Muslim lowlanders began gathering around Hamid Idris Awate, and in September 1961 they established the Eritrean Liberation Front (ELF) and launched an insurgency, in reaction to which Ethiopia officially annexed Eritrea only a year later.

However, Awate never managed to overcome countless issues of ethnicity, clans and ideology between his followers. Furthermore, the ELF found it hard to draw support outside the lowland areas, dominated by Muslims, and the insurgency thus remained rather limited in scope and success for most of the 1960s. By 1970, feeling let down by the British, UN and Ethiopians, many of the younger ELF members began to radicalise and then decided to split and establish their own organisation, the Eritrean People's Liberation Front (EPLF). With the help of money and arms provided by Iraq and Libya, the ELF and the EPLF gradually began to expand their ranks.

Ethiopian authorities initially controlled the situation with police alone. Even after the army deployed two brigades, the 25th and the 33rd of the 2nd Infantry Division, for the protection of Asmara, Akordat, Teseney and the port of Asseb, while the air force deployed around a dozen SAAB B.17s to Asmara, during 1970 the primary

36 Ahenafi, interview, 2001.

The 3rd COIN Squadron IEAF was equipped with eight T-28D Trojans, one of which can be seen here. Notable are underwing gondolas with 12.7mm machine guns. (Photo by Dave Becker)

Thanks to money and equipment provided by several Arab countries, the EPLF insurgents were always well trained and equipped. This photograph shows a group of recruits while undergoing basic training in the mid-1970s. (Albert Grandolini Collection)

responsibility for COIN operations remained in the hands of police. The situation experienced its first escalation on 21 November 1970, when the EPLF launched a campaign, ambushing Ethiopian officers and assassinated the CO of 2nd Infantry Division, Gen Teshome Ergete. Within the following weeks, another insurgency erupted in the state of Tigray and the army eventually felt forced to deploy the entire 9th Mechanised Division in these two areas. The IEAF followed in fashion by deploying two Canberra bombers and at least two F-86s to Asmara. These were soon sighted flying intensive operations against Eritrean insurgents, frequently using napalm bombs.

Furthermore, the IEAF was prompted to establish the 3rd COIN Squadron, equipped with eight T-28As, all of which were upgraded to T-28D Trojan standard before delivery from the USA. Throughout its existence, this unit benefited immensely from cooperation with the US MAAG team, many members of which completed at least one tour of duty in Vietnam. The 3rd Squadron was deployed at Asmara in 1971 and its pilots began flying SAAB B.17s in addition to T-28s.

It did not take long until the activity of the IEAF prompted the EPLF to start claiming Ethiopian combat aircraft as shot down. Amongst others, the insurgents claimed seven aircraft, including two F-86s as downed by 1975, but it seems that it was mostly Trojans that were actually lost, and then primarily to reasons other than ground fire. Namely, not only that the operational standards and combat efficiency of IEAF pilots remained exceptionally high, but the T-28s began suffering engine-related problems caused by intensive flying. Indeed, the 3rd COIN Squadron is known to have been left with only 6 operational Trojans and less than 10 SAABs by 1974.[37]

Nixon Doctrine

Meanwhile, in May 1969, new US President Richard Nixon put forth the essence of what became known as the 'Nixon Doctrine', according to which the USA would assist in the defence of its allies, but would not undertake all the defence of the same, because allied nations were to be in charge of their own security and approach the US defence sector in order to purchase the necessary equipment. While aiming primarily at 'Vietnamization of the Vietnam War', i.e. bolstering defence capabilities of South Vietnam in order to enable US forces to disengage, following this doctrine, the USA vastly expanded its military aid, as well as exports of arms to a number of its allies elsewhere, including Ethiopia. Although experiencing a minor 'dent' following Ethiopian acquisition of Canberras, the value of US assistance provided to Addis Ababa reached about US$10 million annually in 1970, and no less than 25,000 Ethiopian officers and other ranks went through different courses in the USA. Included was an additional group of F-86-pilots – including Heile Michal Biru and Ashenafi Gebre Tsadik the then CO of 1st Fighter-Interceptor Squadron, that converted to F-5s.

Furthermore, the IEAF units equipped with T-28Ds, F-86s and C-47s became involved in a major joint Ethiopian–US exercise codenamed 'Sandy Road', which was simulating the IEAF's involvement in a counterinsurgency scenario and including T-28-operations by day and night (with help from flares). Undertaken in an area north of Dire Dawa, this exercise not only proved highly successful, but also satisfied US instructors, who were very pleased about the air force's ability to absorb new technology and equipment.

Unsurprisingly considering good cooperation, in October 1970, Addis Ababa and Washington signed several new agreements in the framework of which the USA agreed to expand and train an Ethiopian military of 40,000 men in exchange for major expansion of the Kagnew station. By 1973, the value of additional US military stocks furnished to Ethiopia reached US$20 million.[38] The air force was reinforced through the arrival of nine Fairchild C-119K Boxcar transports and, in 1972, two T-33As previously operated by the Royal Netherlands Air Force.[39] The following year, the Pentagon and State Department mediated in negotiations between Ethiopia, Iran and Northrop, resulting in acquisition of three newly-built F-5As and two F-5Bs from the USA, as well as at least seven 'second-hand' F-5As and one F-5B from Iran, deliveries of which lasted through 1973 and 1974. Finally, the Pentagon furnished six ex-US Air National Guard and ten ex-Imperial Iranian Air Force F-86Fs to the IEAF, enabling it to establish the 4th Fighter-Interceptor Squadron.[40]

The cooperation with the USA meanwhile reached such proportions that practically all of Ethiopia's pilots and ground personnel had visited training courses of one kind or the other in the US. For example, when the IEAF received a (very old) early warning radar of an unknown type from the USA in 1973, which was to support operations of its interceptors, Ethiopian personnel were trained to operate this system at Tyndall AFB, in Florida, and

37 Ibid., & Flintham, pp.146–147.
38 Dupuy et al., p. 256.
39 The aircraft in question used to wear serials 'M-11' (IEAF serial 213) and 'M-58' (IEAF serial 214), and were delivered to Debre Zeit on 2 Oct. 1972.
40 Techane, interview, 2006; Techane was one of the IEAF pilots to test-fly F-86Fs transferred to Ethiopia from Iran. He recalled that ex-Iranian Sabres were flown from Dezful via Riyadh to Debre Zeit. Later on, Techane was re-assigned to 5th Squadron, where he flew former Iranian F-5As.

Rare photograph of one of the ex-Iranian F-86Fs seen taking off from an airfield in Saudi Arabia during a delivery flight to Debre Zeit AB in 1974. (Tom Cooper Collection)

The IEAF received nine Fairchild C-119K Packer transports, starting in 1971. They remained in service until replaced by Antonov An-12s, in the early 1980s. (Albert Grandolini Collection)

One of two ex-Royal Netherlands Air Force T-33As, delivered to Ethiopia in 1972. (Photo by Dave Becker)

at Ellington AFB, near Houston, Texas. Together with increasing oil prices, such experiences prompted the commanders of the Ethiopian air force to close all domestic intermediate and jet training facilities, because it proved cheaper to send cadets to the US instead of training them locally. In late 1973 and early 1974, the last group of IEAF pilots trained in Ethiopia concluded their syllabus by qualifying on T-33s and then travelled to the USA for training on Cessna T-37 Tweets and Northrop T-38 Talons, before converting to F-5s. Subsequently, only basic flying training, consisting of around 200 hours on SAAB Safirs and T-28s, was still run at home.[41] At about the same time, the IEAF and MAAG developed plans for constructing comprehensive workshops at Debre Zeit, enabling complex overhauls of all available US-made aircraft and helicopters. Work on corresponding facilities proceeded slowly, and was it was only partially complete before the break of Ethiopian–US relations in 1976.

Eventually, by 1974, the ties between Addis Ababa and Washington were close enough for the IEAF to, following a very careful assessment, request supply of McDonnell Douglas F-4 Phantom II fighter-bombers. Namely, leading commanders of the air force knew all too well that Canberras were acquired in a rush and to counter Somali acquisition of Il-28s, but did not really fit into the IEAF's doctrine of having a small, well-rounded tactical air force. The F-4 could carry an even heavier load of armament than the Canberras, while offering a range sufficient to reach central and southern Somalia, including Mogadishu, while operating out of Debre Zeit.[42]

The US representatives turned this request down, but agreed to provide sufficient Foreign Military Sales credits for acquisition of 16 or 17 Northrop F-5E Tiger IIs, 12 Cessna A-37B Dragonflies, and 15 Cessna 310s instead. Nevertheless, in the course of relevant negotiations, the Ethiopians managed to tie the deliveries of F-5s to equipment required to set up a fully integrated air defence system, supported by an early warning radar network based on several Westinghouse AN/TPS-43D radars. Although the delivery of such systems was initially postponed, primarily because of an Iranian order but also because of subsequent developments in Ethiopia, at least two of these did eventually reach Ethiopia.[43]

Imperial Ethiopian Army Aviation

Another result of the Nixon Doctrine was intensified build-up of the Imperial Ethiopian Army Aviation (IEAA). This little-known branch was originally established at Debre Zeit in 1967 under the command of Col Datchew and his deputy, Col Yigzaw Yimenu. Initially based at a small compound outside the air base, the IEAA moved to Lideta AB, near Addis Ababa a year later, at the same time the Ethiopian Airlines were relocated to the Bole International Airport (originally constructed in 1950, but vastly expanded in period 1961–1963).

The original core of the IEAA consisted of IEAF pilots and technicians, but the first group of different ranks recruited specifically to serve with this branch were largely cadets drawn from the Military Academy at Harar. The pilots received a ten-month course in the USA, structured in three phases as follows:

Phase I: Four months at Port Stewart, Georgia, for 110 hours of training on Cessna T-41 Mescaleros (a military version of the Cessna 172 Skyhawk), including exercises in basic and formation flying.

Phase II: Three months at Fort Rocker, Alabama, for 60 hours of training on Beech T-42 Cochises (a military version of the Beechcraft B-55B Baron), including exercises in navigation, long range and instrument flying.

Phase III: Two months at Fort Rocker, for 50 hours of tactical training on Cessna L-19s.

Upon their return to Ethiopia, most new officers were given a short orientation and operational conversion course to adapt to the local environment, with several MAAG officers, all fresh from a tour of duty in Vietnam, acting as mentors. Additional groups of IEAA students were sent to the USA for training during the following years as well, until in 1973 the decision was taken to continue all further training locally.

The IEAA was initially equipped with four T-41Ds and ten Cessna 185 Skywagons, and their crews underwent the same training curriculum as the original group, in the USA. These aircraft

41 Interviews with several former IEAF officers, provided on condition of anonymity.

42 According to recollections of former IEAF officers involved in related studies, it was for the purpose of becoming capable of launching long-range operations against targets deep inside Somalia, that the airstrip in Gode was expanded into a small airfield with a metalled runway, 'should the need arise'.

43 Letter by Assistant Secretary of Defence, Robert Ellsworth, to George Vest, Director of Bureau of Politico-Military Affairs at Department of State, I-25667/74, dated 30 Oct. 1974, provided in response to FOIA inquiry.

One of ten Cessna 185 Skywagons delivered to the IEAA. This plane was photographed after its evacuation to Djibouti in 1991, which is why it shows the national insignia as in use by Ethiopian military aircraft from 1977 until 1991. (Herve Desallier via Albert Grandolini)

The IEAAF also operated four DHC-6 Twin Otters, delivered in the late 1960s. These light transports could be operated from most primitive runways, and were camouflaged in a very unusual pattern consisting of dark sand and light olive green. (Albert Grandolini Collection)

and their crews were deployed for forward observation and artillery spotting, scouting and reconnaissance for ground units, but also for liaison and transport purposes. Starting in October 1967, the IEAA received the first out of an eventual sixteen Bell UH-1H helicopters, followed by at least six de Havilland Canada DHC-3 Otters and four DHC-6 Twin Otters. The first group of Ethiopians trained on UH-1s in the USA included four pilots, amongst them Maj Tadesse Gebre Tsadik who, upon return to Ethiopia, was appointed the CO of the squadron flying UH-1s. Due to a shortage of fuel as a result of the Arab–Israeli war in October 1973, Tadesse's crew flew only 20 hours during the following year, and did most of their training on simulators.[44]

The five Aérospatiale SA.316B Alouette III helicopters purchased by Ethiopia from France in 1970, entered service with the IEAF instead of the army aviation, but during the following years ever more attention was paid to the new branch, and in 1976 its officers were some of the last Ethiopians to undergo training in the USA.[45]

Barre's Climb to Power

In the meantime, Somalia continued expanding its military. In 1969, the SNDF was reinforced through deliveries of 150 armoured vehicles of Soviet origin, including 100 T-54 MBTs which entered service with four newly-established tank battalions. Additional deliveries followed, including quantities of infantry weapons;

Stylised poster of Maj Gen Siad Barre, who came to power following the coup of 15 October 1969. (Mark Lepko Collection)

BTR-40s, BTR-50s, BTR-152 APCs, anti-aircraft and field artillery pieces to equip nine mechanised infantry battalions, one commando battalion, two field artillery battalions with 76mm guns and 122mm howitzers; two heavy anti-aircraft battalions with KS-19 radar controlled 100mm anti-aircraft guns and three light anti-aircraft battalions with 37mm M1939s and 14.5mm ZPU heavy machine guns.

Furthermore, Egypt supplied twelve different torpedo boats; Sudan trained staff officers, cadets, signal and engineer NCOs; and other Somali ranks were trained in the USSR, People's Republic of China, Egypt, Italy, Iraq and Syria; while the USA, Italy and West Germany supplied equipment and training for the police and a commando battalion until all Western assistance was suspended in 1970. The SNDF thus grew to a total of about 13,000 officers and other ranks, yet its combat effectiveness was still low, primarily due to the lack of maintenance facilities.[46]

Nevertheless, dissatisfied by what they saw as 'slow' development, the SNDF and the police seized power in Mogadishu in a coup that saw the assassination of President Abdirashid Ali Shermarke by one of his bodyguards, and the arrest of the Prime Minister, on 15 October 1969. While dissolving the National Assembly and Cabinet, abolishing all political parties and suspending the constitution, the leaders of the coup established the Supreme Revolutionary Council (SRC) presided over by Major General Siad Barre. Barre immediately announced that his government would support 'all liberation movements in countries under colonial rule',

44 According to recollections of former IEAA pilots, the first 12 UH-1Hs delivered to Ethiopia initially received serials between EA70 and EA82.

45 The IEAF originally ordered two Alouettes and two Aérospatiale SA.330B Pumas. Shortly after the delivery of the first Puma, the IEAF cancelled this order, returned the one delivered Puma and replaced it with two additional Alouettes instead. Two other Alouettes were acquired from Romania in the 1980s.

46 Dupuy et al., pp. 255–256; Brent, p. 143 & Taylor, p. 134.

as well as those in 'illegally occupied territories', obviously referring to the one million Somalis living in Ethiopia, Kenya, and the French Territory of the Afars and the Issas. Although proclaiming a policy of non-alignment in foreign relations, he then requested aid from Moscow and in return for corresponding promises declared Somalia a 'socialist' and 'democratic republic' (as 'al-Jamhouriya as-Somaliya al-Democradiya'), practically turning it into another battlefield of the Cold War.[47] This new official designation was nothing but a farce. The country was ruled by the military that controlled an alliance of several major Somali tribes, kept together primarily thanks to Barre's promises of establishing 'Greater Somalia' which was to include the entire Ogaden, the French Territory of the Afars and the Issas, and much of eastern Kenya, as soon as possible. The Soviets probably understood this, as there was little doubt that there was no other way around realising such plans but by force. Nevertheless, their military assistance increased considerably during the following years, in turn enabling Somalia to position itself on a direct collision course with Ethiopia.[48]

Two youthful future Somali MiG-21 pilots seen in the USSR. (Albert Grandolini Collection)

Somali MiG-21s

As of 1970, the CCS had 1,500 officers and other ranks and was organised into four fighter squadrons operating around 40 MiG-15s and MiG-17s, as well as three other units with up to 20 transports, helicopters and trainers.[49] Readiness rates were still low and most of the fighter pilots flew barely 20 hours a year. The situation was soon to change, though.

Namely, in 1972, Moscow launched a new diplomatic offensive aimed at improving its position in Africa. Soviet Defence Minister Marshal Andrey Gretchko visited Mogadishu and signed an agreement envisaging the improvement and modernisation of the port of Berbera, in exchange for basing rights of Soviet Navy's warships at that facility. This port was important to Moscow because it could offer its navy a nearly permanent presence in the strategically important corner of the Indian Ocean, and an opportunity to counter US Navy operations in the Arabian and Persian Gulfs. During the following years, the Soviets developed Berbera into a major base that included an extensive missile storage facility, an airfield with runways longer than 5,000m and an extensive radar and communication suite. Only two years later, Somalia and the Soviet Union signed a Treaty of Friendship and Cooperation, and the USSR then launched a major effort to modernise and expand the SNDF and the CCS.

Indeed, the Somali Air Force was to yield some of the biggest benefits of this deal. Starting in July 1974, it was reinforced through deliveries of 33 MiG-21MF interceptors and three MiG-21UM two-seater conversion trainers, as well as four air defence battalions equipped with SA-2 and SA-3 SAM-sites.[50]

Interestingly, the CCS decided to leave its organisational structure as it was, and instead organised its new MiGs into two newly-established interceptor squadrons. Except for a handful of commanding officers, the existing MiG-17-units, which were now relegated to fighter-bomber tasks, did not share their experienced pilots and ground personnel with the newly-raised squadrons. Instead, the majority of around 30 pilots and a similar number of ground crew were fresh from jet conversion courses at Primorsko-Atharsk and Bataysk, in the USSR, and thus had very little experience in flying fast jets, and no combat training. However, it seems that their Soviet advisers concluded that any possible problems could be easily solved. The novices were sorted out during an additional course on MiG-21UMs and MiG-21PFMs run at Krasnodar, and then sent back to Somalia where they underwent conversion to MiG-21MFs. Even then, none of the new pilots, nor any of their commanders, were sent to the Air Defence School in Armavir (at least not before 1977), and all of their minimal combat training was undertaken in Somalia. Surprisingly, the number of training accidents remained minimal, which was quite astonishing considering that most new Somalia MiG-21 pilots previously did not even possess a bicycle.[51]

47 Dupuy et al., pp. 255–256.

48 Some sources indicate that one of the strongest arguments for the increase in Soviet military aid to Mogadishu was the discovery of what might be the world's largest deposits of uranium ore in Somalia, in 1968. The finds in question were close to the surface and thus susceptible to economical strip-mining methods. Furthermore, in 1973 the US company ARMCO reportedly made the discovery of oil and natural gas on the Ethiopian side of border of Somalia (for reporting about the later topic, see Bereket H Selassie, 'Conflict and Intervention in the Horn of Africa', *Monthly Press Review*, New York, 1980). However, the author has his doubts about Soviet interest in Somali uranium or Ethiopian oil or gas; despite exploitation of immense natural wealth in many other African countries supported by the Soviet Union in the 1950 to 1970s, Moscow never made any attempts to yield profit from local resources, or even exercise any kind of influence upon their exploitation.

49 Dupuy et al., p. 256 & Brent, p. 143. Note that according to Babich, 590 CCS officers, pilots and technicians underwent training in the USSR between 1963 and 1977. Somali sources cite no fewer than 2,400 as trained in the USSR and another 150 in other countries of Eastern Europe.

50 Figures for the number of MiG-21s were provided by former CCS pilots in interviews granted on condition of anonymity. According to US personnel who served in Ethiopia during the early 1970s, deliveries of MiG-21s to Somalia lasted at least until early 1975, when three aircraft arrived in Mogadishu on board the merchant SS *Akademik Shimansfiy*. Furthermore, the CCS should have received a number of additional 'MiG-15 fighters', in April 1973 too. Such deliveries remain unconfirmed, or were related to deliveries of additional MiG-17s. Detail about deliveries of SA-3s is from *Aerospace Encyclopedia of World Air Forces*, p. 146.

51 Interviews with former CCS pilots and with Robert Szombaty, retired MiG-21-technician of the Hungarian Air Force and expert researcher about former Warsaw Pact air forces, in June 2009. According to same sources, the training curriculum of CCS MiG-21-pilots in the USSR remained the same until at least 1977, when one group of them began flying MiG-21UMs and MiG-21PFMs immediately after their jet-conversion courses, i.e. during their second year in the Soviet Union. The same group of Somalis appears to have been the first to get at least a few hours on MiG-21MFs before returning to Somalia, after three years of training in the USSR.

Ahmed Sheikh Hassan was one of the Somali pilots trained in Iraq in the early 1960s. This photograph was taken when he was decorated for his achievements and service by C-in-C CCS, Col Yusuf Hassan Ibrahim 'Yusuf Deg' on 15 February 1973. Hassan was subsequently appointed the CO of the Flying School in Hargheisa. (Sheikh Hassan Collection)

However, subsequent training of CCS pilots experienced unexpected, but significant difficulties. Although Moscow had increased the number of its advisers in the country from 450 to around 1,500, the number of those assigned to Somali MiG-21-units remained the same. Most new advisers were appointed primarily to the army and military intelligence where their appearance in increasing numbers caused some dissent.[52] Furthermore, in 1974 and 1975, Somalia experienced severe drought, the economy stagnated and thus there was not enough money to buy the necessary fuel. Correspondingly, the number of hours CCS pilots spent in the air during this period increased only marginally, to around 40 annually. As could be expected, the air force soon began experiencing a higher accident rate. Two MiG-21MFs were written off relatively soon after their delivery in the course of a landing accident at Baidoa AB, and a third was written off under similar circumstances in Berbera. Several MiG-17s were written off as well.

The CCS thus entered the critical period during the mid-1970s reinforced in numbers, and supposedly with a much improved combat effectiveness, but actually with the majority of its pilots lacking recent flying experience and without any kind of combat training.

The Fall of Solomonic Dynasty

Significant attempts at developing the economy and modernising local institutions on the part of the ageing Ethiopian Emperor Selassie, during the 1960s and early 1970s, transpired to have been too slow for many groups within the population of Ethiopia, but also too rapid for others. Several of Selassie's attempts at land reform and tax changes failed, causing not only student protests and peasant revolts, but also mounting inflation, corruption and even famine. Facing critique for heavy spending for a relatively large professional military – although many Ethiopians openly expressed pride at their armed forces, seeing them as a guarantee of national independence – the authorities reacted with repression. By 1974, Ethiopia was full of uproar, strikes and demonstrations, and after elements of the army mutinied, central government lost control over several federal states.

The reaction of top military ranks was the establishment of the 120-man Provisional Military Administrative Committee, better known as the 'Derg' (sometimes spelt 'Dergue'), in early June 1974. Three months later Emperor Haile Selassie was overthrown in a bloodless coup d'état, in the course of which the Emperor, his family and all officials were arrested. Organised from within, this coup was followed by a period of confusion in which several different groups began struggling for power. Initially, none of the involved parties had a clear idea about how to organise and run one of world's poorest nations, but eventually the military managed to crush the resistance of feudal Amharian landlords and established themselves in power. The Derg then returned Gen Aman Mikael Andom from retirement and appointed him as the head of state. During a major conference of the Derg on 17 November 1974, several officers demanded the execution of Emperor's officials. Aman, who had known many of detainees, not only because he was educated at the Royal Military Academy Sandhurst in the UK together with Lt Gen Abiy Legess, son-in-law of the Emperor, refused these requests and stormed out of the meeting. Most of the Derg then turned against him; Aman, two former defence ministers and their bodyguards were all killed 'while resisting arrest', during the night of 23 November, followed by 59 members of the royal family and ministers. In their place two young officers, Col Mengistu Haile Mariam and Lt Col Atnafu Abate, established themselves in power.

Thus ended the Solomonic Dynasty that, according to its own legend, ruled over Ethiopia for 3,000 years.

Tiger Deal

Surprisingly, the chaos in the country and the bitter power struggle within the military had only limited impacts upon the IEAF. Some unrest within the air force was reported in mid-February 1974, but General Asssefa Ayene subsequently reported to US instructors that there was no mutiny, rather a strike over various demands, including better wages and working conditions. Certainly, the air force lost the prefix 'Imperial' following the fall of Emperor Selassie. However, only very few of its senior officers involved in the politics, were arrested or forced to leave. Even fewer were killed or executed. Indeed, the most severe blow the EtAF experienced was when two pilots of the Bomber Squadron quit their service; Bezuwork has left on his own, while Afeworki Mekonnen went as far as to defect, flying a Canberra B.Mk. 52 bomber (serial number 354) to Hargheisa AB in Somalia. Not interested in returning this precious aircraft to Ethiopia, the Somalis made Afeworki fly it to Kismayo, where it was left to rot ever since. The sole bomber asset of the air force was thus left with only two pilots and three operational aircraft by early 1977.[53]

Otherwise, the EtAF managed to keep ties to the US military mission in Addis Ababa even after the USA ceased supplying military aid to Ethiopia immediately after the coup of 1974. Such ties were soon used by the new government in attempt to encourage

52 Former US personnel that used to serve in Ethiopia in early 1970, interview provided on condition of anonymity, and citing reports by Italian Military Attaché in Mogadishu from April 1974. According to several interviewed Somali officers, concerned about increasing Soviet presence and influence in Mogadishu, Iran and Saudi Arabia then offered a US$75 million aid package to Barre's government, conditioned on reduction of Soviet activities. However, the Somali strongman rejected this condition and this offer was withdrawn.

53 Afeworki was subsequently released into exile in Canada, where he has lived since 1980 or 1981. His defection was not the first of its kind in this part of Africa. Several years earlier, a CCS pilot defected to Ethiopia flying a Cessna training aircraft. Although the two countries were not in a state of war at that time, that aircraft was also not returned but put into service with the Ethiopian Air Force.

Row of F-5Es waiting for delivery to Ethiopia. They were the centrepiece of the last ever major arms deal negotiated between Addis Ababa and Washington. (Tom Cooper Collection)

the Americans to start providing military aid again. Keen to secure its interests in the Horn of Africa, Washington was ready to listen, and following a series of meetings during late January and early February 1975, the State Department agreed to not only re-start, but also increase its Foreign Military Sales credits to Ethiopia from US$8.3 million to between US$20 and 25 million in 1976. Unsurprisingly, the then Foreign Minister of Ethiopia, Kifle Wedajo, rushed to request 16 Northrop F-5E Tiger II fighter-bombers and AIM-9 Sidewinder air-to-air missiles (AAMs), in addition to 11 AN/TPS-43D early warning radars, and unknown quantities of M60 MBTs, M113 APCs, M109 self-propelled howitzers (SPHs), M163 Vulcan self-propelled anti-aircraft guns (SPAAGs) and BGM-71 TOW anti-tank missiles. Initially, the Americans demanded cash payments for some of these items, which Ethiopia could simply not afford. Nevertheless, a solution was eventually found and later in 1975, the administration of US President Carter authorised the transfer of the first eight F-5Es together with stock of AIM-9B Sidewinders, and eleven M60 MBTs (in addition to 22 that were delivered to Ethiopia at earlier times).[54]

Contrary to earlier times, the process of introduction of F-5Es to service with the EtAF was organised entirely by Ethiopians, although some help was provided by a USAF MIT unit that remained at Debre Zeit for a short period of time. The first six EtAF pilots and a similar number of technical officers underwent a four weeks language course at Lackland AFB, near San Antonio in Texas, starting in August 1975, primarily in order to refresh their 'American English' and re-learn technical terms for flying. Included in this group were experienced F-5A-pilots, including Mengistu Kasaye, Lagesse Tefera, Ashenafi Gebre Tsadik and Girma Workagexehu, as well as two former F-86-pilots, Bacha Hunde and Bezabih Petros. Next, they underwent two weeks of physiological training at Randolph AFB, and parachute training at Brooks AFB. Former F-5A-pilots were then sent to Williams AFB, for a three and a half month conversion course on F-5Es, including lessons in air combat manoeuvring, totalling 48 hours. Once there, the Ethiopians experienced that the early 1970s were the times of 're-discovery' of dogfighting as a viable air combat technique and, unsurprisingly, their training included an Advanced Combat Manoeuvring course provided by USAF pilots. Berhanu recalled:

Four of six Ethiopian pilots that underwent the F-5E-conversion course in Arizona, in 1976, seen together with their USAF instructors (from left): Mengistu Kasaye (kneeling), Lagesse Tefera, Girma Workagexehu, and Ashenafi Gebre Tsadik. (via S.N.)

> After the introduction to the F-5E, the training was exactly the same as on F-5As earlier, and included basic fighter manoeuvring, air combat manoeuvring, air combat tactics and air-to-ground gunnery. Until then, the role of wingman was pretty much to protect his leader only. However, what they taught us in the USA that time was a new role for wingman; he was now to become much more involved, directly cooperating with the formation leader. We trained a lot of weaving manoeuvres to check each other's 'six o'clock' and coordinate attack formations

Ashenafi added:

> They taught us the latest in air combat tactics. The course did not include live missile firing but we did air-to-air gunnery on a target-dart and we flew many simulated dogfights.

While Ashenafi and Girma returned to Ethiopia immediately after the end of this course, Lagesse and Mengistu remained in the USA ten months longer in order to complete an instructor-pilot course. Upon their return in 1976, they trained the first two Ethiopian F-5E-pilots to convert to the new variant at Debre Zeit, Berhanu Wubneh and Afework Kidanu. Girma, on the other hand, became involved in politics and was killed during the subsequent power struggle.

Ultimately, enough EtAF pilots were trained to establish the 9th Fighter-Interceptor Squadron (also known as 'Tiger II Squadron'), which operated only eight F-5Es, delivered by US pilots to Debre Zeit in late 1975, but no F-5F two-seaters; the single-seaters remained the only examples of this variant to ever enter service in Ethiopia.

Apart from F-5Es in early 1976, the EtAF also received its first two AN/TPS-43D radars. One of these was installed high on a

54 *Briefing Memorandum* 7502362, Department of State, dated 3 Feb. 1975, provided in response to FOIA inquiry. The USA eventually delivered a third batch of M60 MBTs to Ethiopia as well, thus bringing the total to 47. Notable is that Ethiopia, despite power struggles in Addis Ababa, multiple insurgencies and an economy in ruins, eventually paid for all these weapons and even made an advanced payment for additional ones, though these were never delivered. In 1977, Washington froze all the Ethiopian assets in banks in the USA and returned these only in 1991.

mountain near the Karamara Pass in the Jijiga area, and the other at Debre Zeit AB. Two groups of Ethiopian officers and other ranks, each consisting of several radar technicians, electricians, controllers and intercept officers, were trained at an unknown base in the US federal state of Mississippi and McLennan AFB in California. The second group had to cut their training short after one month and return early, due to deteriorating relations between Addis Ababa and Washington, but also the threat of Somali invasion.[55]

55 Involved 'Radar-intercept-officers' were trained separately from the rest of these two groups, but no details about their courses in the USA are known. Most of the people in question are still living in Ethiopia and have proved outside the reach of the author.

Overall, the acquisition process of F-5Es and AN/TPS-43s, two systems that were to prove of crucial importance for the flow of the Ogaden War, ended early without all the equipment ordered being delivered although paid for by Ethiopia. Because of the Derg's poor human rights-record, President Carter embargoed the remaining six F-5Es (as well as deliveries of plenty of other aircraft, armoured vehicles and equipment) and ordered the departure of all US military personnel from Addis Ababa in March 1977.[56]

56 The remaining six F-5Es manufactured for Ethiopia were instead delivered to the USAF's 3rd TFW, forming the 26th Tactical Fighter Training Squadron, an 'Aggressor' outfit, at Clark AFB, in the Philippines.

CHAPTER 3
MILITARY BALANCE AND PLANNING

EtAF – Left on its Own

The EtAF was cut off from US support and left on its own at a time when Ethiopia was going through especially bitter and chaotic struggles and when a major crisis was about to develop in relations to Somalia. Nevertheless, the Americans had trained the Ethiopians very well over the previous 16 years, and made especially their air force self-reliant. Except for training, also important was the fact that the USA delivered enough spare parts for F-5As and F-5Es to keep them operational for several years, and that the Ethiopians not only learned how to maintain and operate these aircraft, but also how to deploy them in combat and how to obtain additional spares on the black market.

As of 1976, the EtAF, then under command of Gen Taye Tilahun, had about 3,000 officers and other ranks, organised as listed in Table 1.[57] Notable is that the air force was practically grouped into three wing commands, one set up at Debre Zeit, one at Asmara and one at Dire Dawa. Not only two combat units but also two training units were deactivated in the years before or early on during the Ogaden War. Similarly, despite this organisational structure, the F-5Es of the 9th Squadron had spent most of the first few days of the war deployed forward at Dire Dawa. In fact, despite a number of Russian reports to the contrary, all the remaining F-86s and T-33As were withdrawn from this part of Ethiopia and concentrated at Asmara, where they were to remain for the duration of the Ogaden War. Something similar can be said about the T-28Ds of the 3rd COIN Squadron; while originally deployed at Dire Dawa and even flying a number of combat sorties early during the Somali insurgency in Ogaden, this unit was eventually deactivated and its aircraft stored about two months into the war.[58]

Although such decisions might appear to have limited resources available to the EtAF during the critical point in the nation's history, they eased the planning of operations. The older jets and T-28Ds were shorter-ranged than the F-5s as well as being potentially vulnerable to interception by Somali MiGs; also considering the size of the battlefield, endurance in combination with war load, which older aircraft were lacking, were to prove of crucial importance.

When it comes to numbers of operational aircraft, the 9th Squadron had only seven operational Tiger IIs because one was damaged beyond repair during a nocturnal rocket attack of Somali insurgents against Dire Dawa AB, at an unknown date in early 1977. This action eventually prompted the EtAF to withdraw all of its operational aircraft from that air base to Debre Zeit, every evening. At least two F-5Es and two F-5As would be brought back the next morning, and then withdrawn in the evening again. Only a forward Operations Command Post remained at Dire Dawa, with Techane Mesfin, Amha Desta and several other senior officers rotating as commanders responsible for assigning specific missions to forward deployed pilots.[59]

Ethiopian Air Force's F-5Es were to bear the brunt of combat operations during the first six months of the Ogaden War. (Albert Grandolini Collection)

The 5th Squadron is known to have had a total of eighteen aircraft and fifteen pilots as of July 1977, including two F-5Bs, and one Freedom Fighter that was modified to RF-5A standard through addition of suitably equipped nose section with several reconnaissance cameras. The single RF-5A was primarily flown by Tesfu Desta, although Lagesse Teferra flew a few operational sorties with it as well. The serviceability of F-5As and F-5Es varied from day to day depending on the intensity of operations, availability of spares and combat damage, but generally the EtAF found the

57 Taye later became the Minister of Defence before defecting, in the late 1980s. He is currently residing in the USA.

58 For Russian reports about deployment of Ethiopian F-86s and T-33s in combat against Somalia, see Babich, *With Foreigners against Our Own*.

59 Ashenafi, interview, 2001 & Berhanu Wubneh, interview, 2006.

Table 1: EtAF Order of Battle, mid-1977

Unit	Base	Equipment	Notes
1st Fighter-Bomber Squadron	Asmara	16 F-86Fs	CO unknown
3rd COIN Squadron	Dire Dawa	6 T-28Ds 12 SAAB B.17	CO unknown; unit deactivated in late summer 1977 and aircraft stored at Dire Dawa
4th Fighter-Bomber Squadron	Asmara	16 F-86F	CO unknown; unit deactivated in 1975 or 1976
5th Fighter Squadron 'Tiger Squadron'	Debre Zeit	13-18 F-5A/Bs	CO Lt Col Techane Mesfin; Maj Berhanu Kebede took over during the Ogaden War
9th Fighter Squadron 'Tiger II' Squadron	Dire Dawa	7 F-5Es	CO Lt Col Berhanu Wubneh
14th Helicopter Squadron	Debre Zeit	3 SA.316B	CO unknown
21st Composite Transport Squadron	Debre Zeit	20 C-47 2 C-54 9 C-119K	CO Lt Col Fanta Belay (later C-in-C EtAF)
31st Training Squadron	Dire Dawa	12 Safir	CO unknown; primary training unit deactivated in 1973
32nd Training Squadron	Dire Dawa	T-28A/B	CO unknown; intermediate training unit; aircraft stored at Dire Dawa
33rd Training Squadron	Dire Dawa	T-33As & RT-33As	CO Lt Col Amha Desta; operational conversion unit; re-deployed to Asmara during Ogaden War
44th Bomber Squadron	Debre Zeit	3 Canberra B.Mk.52	CO Major Bizuayehu (Bizuwork)

condition of these two units as 'satisfactory'.[60]

In terms of other EtAF aircraft that were deployed during the Ogaden War, all three remaining Canberra B.Mk. 52s of the 44th Bomber Squadron, as well as most of the transports operated by the 21st Composite Transport Squadron were available, together with SA.316B Alouette III helicopters. Similarly, the Ethiopian Army Aviation (EAA) deployed all of its 16 UH-1H helicopters for troop transport and several Cessnas for liaison and observation during the war.

Ethiopian militiamen on their graduation ceremony at Camp Tatek in the summer of 1977. Large number of militia units were in the process of training and about to be rushed to the battlefields in Eritrea and Ogaden during the second half of the year. (Albert Grandolini Collection)

The main bases of the EtAF and the EAAF as of 1977[61] were Debre Zeit and Dire Dawa, but additional military facilities were available at airports in Asmara, Asseb, Gondar, and Gode, while airfields with unpaved runways existed in Agordat, Jijiga and Mekelle.

Ethiopian Army – in Turmoil

Preoccupied with the continuous power struggle in Addis Ababa and the insurgencies in Eritrea and Tigray, the Ethiopian government was slow to respond to the emerging crisis in Ogaden, even though its top military officers had anticipated a two-stage Somali offensive of the kind that was eventually undertaken in 1977, two years earlier.[62] The situation in Eritrea during the first few months following the coup against Emperor Selassie was especially critical. Asmara, the capital of this state with a population of around 250,000 at that time, nearly fell to combined attacks of ELF and EPLF in January 1975. With hindsight, it can be said that only bickering between the two Eritrean insurgent organisations, the forces of which outnumbered the Ethiopian Army units deployed in the area by three to one, and strong intervention of the air force, prevented the fall of Asmara and perhaps even of the entire Eritrea. Once the T-28s re-deployed from Asmara to Dire Dawa in the summer of 1975, the insurgents became active inside several Eritrean cities again. On 13 September they attacked the US base in Kagnew, killing nine American and Ethiopian soldiers.

The Ethiopian Army (EA) was hopelessly out of condition and unable to do anything against the Somalis: it had only one infantry division and a number of minor units deployed in Ogaden, and most of these were weakened because their elements were re-deployed elsewhere. With no regular army units available to deploy as reserves or reinforcements, in February 1977 the government established the 'Tatek' ('Be Girded') camp on the outskirts of Addis Ababa and started training a large number of 'People's Militia' units, drawn mainly from the peasantry. In a crash programme undertaken with the help of Cuban instructors, during the following months this camp trained as many as 120,000 fighters for no less than ten infantry 'divisions', two of which, the 2nd and the 5th, were dispatched to

60 Berahnu, interview, 2008; except for its CO, Maj Berhanu Kebede, pilots flying F-5As with 5th Fighter Squadron during the Ogaden War were: Alemayehu Gudere, Amha Desta, Asmare Getahun, Berhanu Kebede, Heile Michael Birru, Lulu Gebre Medhin, Mamo Gudeta, Neguissie Zergaw, Seyoum Alemu, Techane Mesfin, Tekele Abebe, Tesfu Desta, Teshale Zewdie, and Wagira Gemechelu. Lt Col Techane Mesfin from 9th Squadron and Tigneh Woldegiorgis, who flew F-86s when the war erupted, flew operations in F-5As as well, and were eventually transferred to 5th Squadron (where they continued serving before joining Ethiopian Airlines, in 1982).

61 The order of battle as provided here remained valid until well into 1978, changes appearing only once the EtAF was nearly completely re-equipped with Soviet-made aircraft.

62 Tareke, p. 643.

Table 2: EA Order of Battle in Ogaden, mid-1977		
Unit	**Headquarters**	
3rd Infantry Division		
5th Infantry Brigade	Dire Dawa	only one battalion still operational
9th Infantry Brigade	Gode	only one battalion still operational
10th Mechanised Brigade	Kebridehar	including 20th Armoured Battalion with 42 M41 Walker Bulldog and M47 Patton tanks
2 artillery battalions	Aroresa (Jijiga)	total of 48 different artillery pieces and 10 Bofors L60 anti-aircraft guns calibre 40mm
11th Infantry Brigade	Degehabur	only one battalion still in Ogaden
24th Nebelbal ('Flame') Brigade	Aysha	
75th Militia Brigade	Aysha	including 752nd Battalion
79th Militia Brigade	Gode	including 219th Nebelbal Battalion
92nd Mechanised Brigade	Jijjiga	re-deployed to Ogaden after the start of war

Ogaden in July 1977.[63] However, when Somalia launched their invasion, the EA only had the relatively few, much weakened units mentioned in Table 2 in Ogaden.[64]

Derg turning East

Attempts by the Derg to impose land reforms and increase government control over local peasantry resulted in widespread discontent and then an outright revolt in 1976. Fighting spread out of Tigray and into southern Ethiopia because Afars, who had a semi-autonomous state ruled from the town of Asahita by Sultan Ali Mireh, opposed the military rulers. All attempts at negotiations failed and Mengistu's and Atnafu's climb to power thus resulted in increasing political chaos and another bloody power struggle in Addis Ababa. The Derg's 'revolution', initially supported by the majority of the population became, due to land-related issues, highly unpopular. Stunned, the Derg split; while the majority of officers supported Mengistu, the rest sided with the former Chief of Staff Army, Brig Gen Teferi Bante, who established himself in power in 1976. However, Bante's government proved unable to bring the situation under control. After Mengistu's followers sided with an extremist Marxist-Leninist group named Meison and led by Haile Fida, Bante declared a war on them, launching the so-called 'White Terror' campaign. The Derg and Meison reacted with an equally brutal 'Red Terror' campaign and this clash provoked fierce firefights all over Addis Ababa. Usually, available estimates claim that over 50,000 of those working or sympathising with Bante's rule were killed during that battle. It culminated in hand-to-hand

63 Gilkes, p. 723.

64 Based on Tareke & Sarin et al., *Allien Wars*; for comparison, Russian sources state the strength of the Ethiopian Army in Ogaden at the start of fighting at around 10,500 troops, 45 M41 and M47 MBTs, 25 T-34/85s, 48 artillery pieces and mortars, and 'around two dozen Bofors 40mm L60 flaks (see Babich, *With Foreigners against Our Own*). However, Ethiopia did not receive any kind of Soviet equipment until several months into the war. Although Yemen did deliver two companies of T-34/85 tanks, these arrived only in Sept. 1977, and did not see any combat service until several weeks later.

M151 MUTT jeep of the Ethiopian Army on the streets of Addis Ababa during the power struggle in 1976. (Albert Grandolini Collection)

Colonel Mengistu Haile Mariam established himself in control of the Derg and thus power in Addis Ababa, following a period of bloody struggle for power. (via S.N.)

fighting in the Grand Palace of the Ethiopian capital, in the course of which Bante was killed, on 3 February 1977. Almost as soon as Bante was dead, the Derg turned against Meison and eliminated its leaders and by the end of the month, Mengistu Haile Mariam established himself in power.[65]

The Derg then decided to solve all its other problems in a similar fashion, and announced their intention to vastly expand the military and crush all the uprisings in Eritrea, Tigray and even the insurgency launched by Somalis in Ogaden. Mengistu's triumph against the Ethiopian Marxists was strangely hailed by the Cuban leader Fidel Castro, and even by the Soviet Union. Correspondingly, Mengistu entered negotiations with Soviet government and in March 1977 proudly announced that he had reached an agreement to obtain Soviet weapons and equipment. No less then thirteen related cooperation agreements and contracts were signed by the Ethiopian leader during his visit in Moscow in May 1977.

While these early Ethiopian–Soviet agreements had next to no impact upon the Ethiopian Army, they did have a clear effect upon the EtAF and the Army Aviation. Namely, one of first orders the Derg placed in Moscow was for twelve MiG-21bis interceptors and twelve Mi-8 helicopters. Correspondingly, the air force had to assemble a group of twelve pilots, led by Haile Michael Birru and including Ashenafi, Berhanu and Teshale, and send these to the USSR for a four-month conversion course on MiGs just as the

65 Thompson et al., *War in Peace*, pp. 248–249.

crisis in Ogaden was about to erupt. Indeed, the pilots in question were just about to complete their theoretical training when Somalia invaded Ogaden, but they remained in the Soviet Union until late August 1977, except for Birru, who was recalled to take over as the CO Dire Dawa AB. Furthermore, the second group of Ethiopian pilots, all of them former F-86-fliers, was also already in the USSR by the time of the Somali invasion.

Tiger II's Advantages

The few F-5E and F-5A pilots not sent to the Soviet Union were meanwhile busy flying intensive dissimilar air combat exercises, pitting Tiger IIs against Freedom Fighters, during which the later simulated the less powerful MiG-17s and MiG-21s.

The exercises in question were primarily based on intelligence about the MiG-21 provided to the EtAF by US advisers. This intelligence assessed the F-5E as clearly superior to MiG-21s in quite a few areas. Foremost, the US-built fighter was capable of much better endurance, which meant that it could reach further, or do so faster, or remain engaged in air combat for a longer period of time. At least as important was that the F-5E did not bleed energy in turning as fast as the MiG-21 and that its engine response to the pilot's input was much quicker. The manoeuvring flaps of the F-5E deployed automatically in air combat, while those on the MiG-21 were actually envisaged as an aid for landing and had to be manually put out. Indeed, if a MiG-21 pilot made a mistake with the set-up of his manoeuvring flaps, the aircraft was likely to depart from controlled flight. Furthermore, not only the cockpit-ergonomics and 'switchology' of the F-5E's systems, but also its much underestimated Emerson APQ-153 radar, were considered superior to those of the MiG-21, and the Tiger II was equipped with a much more advanced gunsight. Eventually, the US pilots, and thus the Ethiopians too, concluded that the MiG-21 might be superior to the F-5E in manoeuvring at low speeds, but as of 1977, only a very few pilots outside Syria (since the mid-1960s) and Egypt (since the early 1970s) have ever attempted to fly their MiGs in such regimes and fully exploit their capabilities in that type of arena.[66]

Because of this, but also because their F-5As lacked the wiring of Sidewinder AAMs, the Ethiopians decided to deploy this variant for ground attack only, while F-5Es were to fight for air supremacy.

Overall, the EtAF thus entered the Ogaden War fairly well-prepared, even though critically short of pilots. However, most important was the fact that, also contrary to many Russian and Somali reports, the Ethiopian Air Force was ready to fight on its own and did not need any foreign advisers. Indeed, no Israelis, Soviets or any other foreign instructors were working with the EtAF, and it was only Ethiopian pilots that flew EtAF fighters during the first six critical months of the war.[67]

66 Interviews with diverse EtAF pilots and with Lt Col Jack Manclark, former CO 4477th Tactical Evaluation Squadron (USAF unit that flew MiG-17s, MiG-21s and other Soviet-made fighters in the 1970s and 1980s), June 2009. Manclark flew 900 sorties in F-5Es of USAF's 'aggressor' squadrons and 301 in MiG-21 in the course of the 'Constant Peg' Programme.

67 It was foremost the Somali media that began reporting about the presence of 'two Israeli squadrons with 24 F-4E Phantom II fighter-bombers' in Ethiopia. It seems that such reports were instigated by Mogadishu in reaction to the stunning success of EtAF pilots in air combats against the CCS' MiGs, in the course of which they admitted to have lost more than twenty jet fighters (*Newsweek* magazine forwarded such claims in one of its editions from September 1977), and supposed 'messages on readiness of Israeli pilots to assist Ethiopia', mentioned by Mezentzev in *Ethiopian–Somali Border War*, p. 36.

Ashenafi Gebre Tsadik, as seen shortly before the war with the F-5E '421', after winning the EtAF annual gunnery trophy. (via S.N.)

Somali Insurgencies in Ogaden

Ethnic Somalis launched an insurgency in Ogaden in early 1970, under the aegis of a separatist organisation that became known as the 'West Somali Liberation Front' (WSLF, originally established in the early 1960s). Through 1976, this insurgency spread throughout Hararghe and into southern Bale and Sidamo, crippling what was left of Ethiopian authority and the economy in the area, and even cutting the country's only railway line, connecting Addis Ababa with the French Territory of Afars and Issas. Eventually, the WSLF grew in size and combat effectiveness to the point where it became capable of not only tackling Ethiopian police and civilian authorities, but also directly attacking units of the Ethiopian Army's 3rd Division, forcing them to abandon their bases. Eventually, the WSLF found itself in control of most of Ogaden and the Bale–Sidamo lowlands by early 1977. Although not managing to take any towns, it put most of these under siege by cutting off communications between them. Correspondingly, during a meeting of WSLF's Central Committee in Mogadishu in the spring 1977, the front publicly claimed all the territory east of the line connecting Moyale, on the Kenyan border, through Awash (some 160km east of Addis Ababa) to the French Territory of the Afars and the Issas for itself. This territory covered almost one third of Ethiopia.[68]

Barre's government had previously been involved in supplying the WSLF with arms and ammunition and officially recognised it as representative of Somalis living in 'Western Somalia', as Ogaden was designated by Mogadishu in early 1975. However, the Somali administration never managed to establish itself in direct control of the WSLF and thus in 1976 the Somali strongman attempted to negotiate regional autonomy for Ogaden by sending his representatives to discuss the issue with the Derg.[69] Once these efforts proved fruitless, during the same year he ordered the establishment of another insurgent group, the Somali-Abo Liberation Front (SALF). Originally, the SALF was led by a few of the WSLF veterans from the early 1960s, but actually it functioned as a proxy of the government in Mogadishu and was primarily

68 Thompson et al., *War in Peace*, pp. 250–251.

69 Nkaisserry, p. 14.

WSLF insurgents with a captured Ethiopian Army jeep in early 1977. (Albert Grandolini Collection)

staffed by SNDF officers and other ranks that resigned their commissions to fight in Ogaden. Indeed, there are reports that over time the Somalis deployed no less than nine of their army's brigades to serve as the core for SALF divisions, each of which was estimated at around 1,000 fighters, and which were closely corresponding to divisions of local Somali clans, both in composition and zone of operations. Some official Ethiopian sources indicate that by July 1977, no less than 39,450 Somali fighters had entered Ethiopia, half of them being deployed to Hararghe and the other half distributed in Bale, Sidamo and Arssi. Another 34,000, mostly equipped with AK-47 assault rifles and machine guns, but also some RPG-7s were to follow, eventually raising the total to 63,200.[70] While the majority of foreign observers have doubts about these figures, and believe that there were never more than 45,000 Somali insurgents inside Ogaden, it is certain that the Ethiopians not only used their military intelligence, but also the sole EtAF RF-5A for the purpose of tracking relevant developments.[71]

Regardless how many fighters the WSLF and SALF might have deployed inside Ogaden, and regardless of them claiming to be in control of 'seven towns' by May 1977, their first major operation proved anything but successful. In late June, the insurgents began assaulting EtAF air bases at Gode and Dire Dawa, and the nearby army barracks. However, these attacks backfired because they exposed the Somalis to the devastating firepower of the defenders and resulted in severe losses. Although they managed to knock out a few aircraft on the ground, as many as 300 insurgents were killed in the attack on Gode alone.[72] Eventually, Mogadishu concluded that an invasion of Ogaden by the SNDF was the only solution.

Somali Motives and Planning

Whether the Somali strongman Siad Barre was born in Shilabo, or in Ogaden and emigrated to Somalia before joining the Italian colonial police in the 1950s, or if he was born in Garbarhaarrey in Somalia (sources differ), it is certain that his mother was from the Ogaden-Darod clan.[73] One of crucial reasons for his rise to power in Mogadishu was his repeated promise of bringing Ogaden under Somali control. The issue of Ogaden was therefore very much one of personal importance for him, not exclusively related to his ethnic and cultural background. Similarly, although many other Somalis made their claim for Ogaden foremostly on the basis of ethnicity, economic interests also lay behind their expansionist intentions. Ogaden includes the Hararghe area with its rolling plains and lush valleys watered by numerous rivers and ample seasonal rains, as well as two major cities, Dire Dawa and Harar. Ethnic Somalis are a distinct minority in this area, known for successful cultivation of teff, barely, wheat and coffee. Almost equally important is the Karamara Pass, that cuts through the mountain range between Harar and Jijiga, and a stretch of railway between Addis Ababa and Djibouti in the area.

Unsurprisingly, the government in Mogadishu carefully monitored developments in Ethiopia during the period 1974 to 1977, and to them the country appeared at its weakest in late 1976 and early 1977, when political chaos and murderous power struggle raged. Excellently informed about the imprisonment and execution of nearly all the top military commanders and the Ethiopian Army's preoccupation with fighting insurgencies in Eritrea and Tigray, the Somalis concluded that an opportune moment for launching an offensive into Ogaden had come. Even then, they acted very carefully, preparing a well-staged strategic campaign. This was to start with the spread of WSLF insurgency, followed by an invasion by the SALF that aimed at the capture of entire Ogaden. Several weeks later, preferably after Barre completed his tenure as the 11th Chairman of the Organisation of African Unity (OAU) in June 1977, a full scale assault by the Somali National Army (SNA) would begin.[74]

The plan for the SNDF's invasion was developed by Maj Gen Mohammed Nur Gallel, Chief of Staff Army and acting commander of the 26th ('Northern') Division. With the majority of the Ethiopian armed forces concentrated in isolated pockets around Jijiga and Gode, preoccupied fighting the WSLF, Gallel could afford to order his forces into the advance without concern for flank protection to positions from where they could launch a classic double envelopment of enemy positions. This advance was to be launched simultaneously from the north and south, so that the force could be re-supplied from depots in Hargheisa, Baidoa and Mogadishu. The principal targets of the first 30 days were Harar and Dire Dawa, followed by the main line of communication between Addis Ababa and Djibouti, the capture of which was expected to seal the fate of Ethiopian control of Ogaden. The operation was to end within 60 days, with all major urban centres and communications in Somali hands. This was a well-conceived plan that had all the early indications of success, because it offered the massed SNA forces a distinct military advantage over the Ethiopians.

Somali National Army

The Somali National Army was probably the best equipped and trained part of the SNDF in 1977. Led by officers trained at top military education facilities in Egypt and the USSR, it entered the Ogaden War significantly reinforced. Then through 1975 and 1976 Soviet military aid continued to arrive, enabling a significant expansion of available assets. All four tank battalions, the pride of

70 Tareke, p. 640 (citing Lt Col Kassahun Tirfe, the deputy intelligence officer of the 3rd Infantry Division of the Ethiopian Army and Col Ketema Gabra Mariam, a paracommando officer) & Nkaisserry; other observers believe that actual figures did not exceed 45,000 men.

71 Ashenafi, interview, 2001.

72 Human Rights Watch, *Evil Days*, p. 75, report released in 1991.

73 Mohammad Haji Mukhtar, *Historical Dictionary of Somalia (New Edition)*, (Scarecrow Press, ISBN 978-0-8108-4344-8).

74 Tareke, p. 639.

Somali soldiers, here a company of SNA recruits during training, were considered some of the toughest and best-trained in Africa of the mid-1970s. (Albert Grandolini Collection)

A T-54 MBT of the SNA. Of interest is not only the large turret number, identifying the position of vehicle within the battalion, but also the SNA patch, that was applied on all of its vehicles. (via Sheikh Hassan)

Underlying the coming war as an all-out effort to conquer Ogaden, Minister of Defence of Somalia, Muhammad Ali Samater, went as far as to mobilise all B and even C reserves of the SNDF. (Sheikh Hassan Collection)

the army, were re-equipped with T-54 MBTs and, coupled with four mechanised battalions, four artillery battalions equipped with 122mm D-30 howitzers were used to establish four mechanised brigades. Two additional tank battalions used the older T-34s and primarily supported infantry units, while two newly-established artillery brigades were equipped with BM-21 multiple-rocket-launchers (MLRs). The Somali Army was not only well-trained and equipped, its combat readiness and morale were very high, not only because many of its officers and other ranks had had fresh combat

Table 3: SNA Order of Battle, mid-1977

Unit	Headquarters	
26th (Northern) Division Dire Dawa Front	Hargheisa	CO Maj Gen Mohammed Nur Gallel CO Brig Gen Aidid
14th Armoured Brigade		equipped with T-54 MBTs
?? Armoured Brigade		equipped with T-54 MBTs
17th Motorised Brigade		-
?? Motorised Brigade		-
15th Infantry Brigade		-
18th Infantry Brigade		-
23rd Infantry Brigade		-
?? Artillery Brigade		equipped with BM-21s
?? Artillery Brigade		36 artillery pieces
?? Tank Battalion		equipped with T-34 MBTs
60th Division Gode & Bay Front	Baidoa	CO Col Abdullahi Ahmed Irro
2nd Armoured Brigade		equipped with T-54 MBTs
?? Armoured Brigade		equipped with T-54 MBTs
?? Motorised Brigade		-
?? Motorised Brigade		-
?? Motorised Brigade		-
?? Infantry Brigade		-
?? Infantry Brigade		-
?? Artillery Brigade		36 artillery pieces
?? Tank Battalion		equipped with T-34 MBTs
54th Division Nugaal Front	Garowe	
?? Infantry Brigade ?? Infantry Brigade		
21st Division	Dusa Mareb	
?? Infantry Brigade		

* Note that according to reports by US personnel that used to serve in Ethiopia during the early 1970s, the sole Somali Army T-34-equipped tank battalion based in the Hargheisa area received 25 additional vehicles from the USSR in April 1973, and was subsequently split into two units. However, available Somali orders of battle for the Ogaden War indicate the presence of only one T-34 equipped unit.

experience and possessed useful knowledge of their enemy, the natural environment and excellent intelligence on the Ethiopian order of battle and positions, but also because the Somalis were keen to fight for the realisation of their national dream.[75]

The SNDF deployed not only all of its major regular units along the border with Ethiopia, the Minister of Defence Muhammad Ali Samater went so far as to mobilise the B-reserves, followed immediately by C-reserves.[76] In this way, the army was able to establish a total of four armoured brigades (with some 600 MBTs and APCs), 22 infantry and ten motorised brigades (equipped with around 200 APCs and 100 tanks), three commando and four artillery brigades (with over 250 artillery pieces), and one transport brigade with around 340 heavy vehicles. These forces were organised as detailed in Table 3.[77]

75 Nkaisserry, pp. 14–15.
76 Aidid et al., Chapter 14.
77 This data as well as Table 4 are based on the cross-examination of all available sources. Notable is that the SNA's divisions were also called 'Commands' or 'Fronts', which often received geographic designations depending on their zone of operations. Later during the war, two additional divisions/commands/fronts were established.

Table 4: CCS Order of Battle, mid-1977

Unit	Base	Equipment	Remarks
Hargheisa Wing			
Bomber Squadron	Hargheisa & Berbera	3 Il-28	CO unknown; unit later withdrawn to Berbera
Fighter-Bomber Squadron	Hargheisa	14 MiG-17	CO Capt Abuker Hassan Hussein
Fighter Squadron	Hargheisa	8 MiG-21MF	CO name unknown, rank of Major
Baidoa Wing			
Fighter-Bomber Squadron	Baidoa	11 MiG-17	CO unknown; unit later re-deployed to Hargheisa
Fighter Squadron	Baidoa	8 MiG-21MF	CO name unknown, rank of Col
Transport Squadron	Mogadishu	2 An-241, An-261, C-471, C-45	CO unknown; elements regularly rotated through Berbera
Flying School	Hargheisa	7 MiG-15UTI, 3 MiG-21UM, 13 MiG-21MF, 8 P.148	CO Col Ahmad Sheikh Hassan
Helicopter Squadron	Mogadishu	4 Mi-4s, 8 Mi-8s	CO unknown

Little is known about options and the ability of the C-in-C CCS, Brig Gen Ali Mataan Hashi, to influence the planning and execution of the Somali invasion of Ogaden. It is certain that, like the entire top of the SNDF, he fell for a number of intelligence blunders regarding the EtAF's ability to fight back. (Sheikh Hassan Collection)

Flying Leopards

The entire Somali armed forces were, at least on paper, quantitatively inferior to those of Ethiopia, and the CCS, then under the command of Brig Gen Ali Mataan Hashi, was much smaller than the EtAF. As of 1977, it still totalled only about 1,750 officers and other ranks. However, in terms of aircraft it could deploy over Ogaden, Somalia was quantitatively superior to Ethiopia. Nicknamed the 'Flying Leopards' for a stylised national mascot of Somalia applied on many of their aircraft, the Somalis could draw upon 29 single-seater MiG-21MFs in two squadrons, and at least 25 operational MiG-17s, also flown by two units. The Flying School probably provided the air defence of Mogadishu during the war, while the sole bomber squadron flew three Il-28s from Hargheisa AB.[78]

These assets were organised in two wings, one based at Hargheisa AB, with some reserve airframes at Berbera, while the other was based at Baidoa AB. Each of these operated two small fighter squadrons and their support services, but also included self-contained air defence batteries equipped with 23mm ZU-23s, 37mm M1939s and SA-7 MANPADs. The balance of the air force was concentrated in Mogadishu, as listed in Table 4.[79]

78 Interviews with former CCS pilots based on condition of anonymity. The C-in-C CCS, Brig Gen Ali Mataan Hashi, was imprisoned by Barre's government shortly after the end of the war and released only ten years later. He served as Chairman of Somali Airlines in the late 1980s, before leaving his country and emigrating to Rwanda in 1991, where he still resides. Note that despite reports about re-deployment of Il-28s from Hargheisa to Mogadishu, in April 1973, Hargheisa served as the main base for these aircraft since their delivery, and until their demise during the Ogaden War.

79 This order of battle remained valid not only during the Ogaden War, but also for most of the time until 1989, even though some units were subsequently re-equipped with aircraft obtained from China, Oman and elsewhere. Available Somali and other sources differ strongly in regards to the number of MiG-17s available to the Hargheisa Wing at the start of the Ogaden War. Most reliable figures indicate the presence of fourteen aircraft, of which an average of twelve were operational. Similarly, Babich reports that the CCS operated no less than ten Il-28s as of 1976 to 1977. However, all available Ethiopian and Somali sources deny it, stressing that Somalia never received this many bombers. On the contrary, out of four originally delivered, one crashed well before the war due to bird strike, killing its Somali pilot and a Soviet adviser, while another crashed during a combat sortie east of Jijiga, sometime in July 1977. Therefore, for most of the war the CCS actually operated only two Il-28s.

Following Soviet doctrine, the primary task of the CCS was to support ground operations. Given the small size of the air force, the Somalis decided to mobilise even their Flying School, equipped with MiG-15UTIs, MiG-21UMs and MiG-21MFs, although next to nothing is known about the operations of this unit during the war. Indeed, they even recalled all of the students undergoing training in the USSR, regardless of their qualification, and seem to have rushed cadets in the first year of their training on L-29s into service, although probably not as combat pilots.[80]

Intelligence Blunders

Although preparing their military plans for invasion of Ogaden to very high standards, the Somali military and political leadership made several crucial mistakes. Foremost, they fell for several major intelligence blunders of their own, as well as those of Soviet origin, which convinced them of a number of advantages, including the quantitative and qualitative superiority of their armed forces. These blunders ranged from mistakes in strategic and operational planning, to a failure to obtain realistic intelligence about the readiness of the Ethiopian Air Force. For example, because its Soviet advisers did not hold him back, Barre expected to receive full support from Moscow, and thus failed to recognise that the leaders of the USSR's foremost interest was in maintaining control over the Horn of Africa, and they actually sympathised with quasi-Marxist insurgents in Eritrea. Barre and his aides also failed to predict how the Central Committee of the Communist Party of the Soviet Union (CPSU) would bring its decision about whom to support, once the Somali invasion was launched, and that the CPSU would eventually assess that establishment of good relations with the Derg promised greater benefits than relations with Somalia.[81]

Barre and his generals were fully aware of the fact that their military could not sustain a long war of attrition and that the war had to be brought to a swift conclusion if Somalia was to win. Not only would SNDF troops advancing into Ogaden depend

80 Robert Szombati, interview, June 2009.

81 Mezentzev, *Ethiopian–Somali Border War*, pp. 29–30.

on an uninterrupted flow of supplies from depots in Somalia, but also these depots would soon run dry if not refilled by the Soviets. However, the flow of Soviet military aid that enabled the expansion of the SNDF, prompted the Somalis to underestimate the degree of Moscow's interest in maintaining influence on the Horn and the length to which the Soviets were prepared to go in order to see their interests preserved. Actually, before the war, the Soviets strongly pressed Barre to avoid hostilities and find a peaceful settlement. Indeed, what Somalis failed to understand was that the failure of Soviet mediation and Barre's uncompromising position regarding Ogaden was eventually to make Ethiopia more attractive for Moscow.

Finally, Barre and his aides also overestimated the extent of support they could expect from the West and various Arab nations.

At the operational level, the SNDF's plan for invasion never involved the WSLF. The insurgents were left to their own devices, independently from the SALF and the Somali National Army. In this fashion, the insurgents exhausted their capabilities in unnecessary operations and were not on hand to help the regulars when the need arose.

At the tactical level, convinced the EtAF was in a similar state of chaos as the rest of the Ethiopian military, the Somalis did not expect any serious resistance in the air and thus did not prepare sufficient contingencies for protection of their supply links. Vadim Koshelev, one of the Soviet advisers that served in Somalia during the Ogaden War, later recalled:

> I learnt about the country to which they wanted to send me only after my arrival in the Ministry of Defence in Moscow, as was usual back then. I was to travel to Somalia. Once there, I was assigned to a fighter squadron in Berbera. All the air base facilities were large and distributed very nicely. One could station two or three regiments of combat aircraft there without any problem. Most fascinating was the control tower; six stories high, very spacious and providing a panoramic view of the terrain and airspace around us. There was a Soviet naval base nearby, and Soviet transport aircraft and airliners were landing all the time, which eased the feeling of being cut off from our homeland a bit. The situation there was placid and relaxed, everything fine, nobody expected any kind of trouble. The war with Ethiopia began almost without notice. Very little changed within the unit to which I was assigned and the Il-28-squadron nearby. The tempo and intensity of operations changed barely at all. The Somalis later began flying slightly more but everything remained the way it used to be. I remember talking several times with Somali CO of the air base, proposing to him to construct several blast pens and to disperse aircraft, which were exposed in the open most of the time. He laughed in response, 'this is not a real war and nothing bad is going to happen.[82]'

Except for not expecting any serious resistance from the EtAF, the CCS was convinced that its aircraft were superior to the equipment available to the Ethiopians. Namely, through the spring of 1976, test pilot V N Kondaurov of the Soviet Air Force had flight-tested a F-5E captured by North Vietnam during the fall of South Vietnam in late April and early May 1975, and brought to the USSR. Surprisingly, although providing a number of recommendations for how to manoeuvre the MiG-21 in air combat against the US-made lightweight fighter, he not only concluded that the Tiger II is only

82 Babich, *With Foreigners against Our Own*.

Pre-war photograph of four Ethiopian F-5As. Although massively underestimated by Somalis, these lightweight fighter-bombers were flown by excellently trained pilots and were to prove decisive for the flow of the Ogaden War. (Craig Kaston Collection)

a 'roughly equal' to the Soviet fighter in terms of manoeuvrability and other major flight characteristics, but that the MiG-21 is faster and in possession of significant advantages due to its radar and R-3S AAMs. At least some parts of Kondaurov's report reached the CCS and Somali MiG-21MF pilots concluded that their only disadvantage to F-5Es was the shorter range of their aircraft, which was barely sufficient for them to reach Dire Dawa when carrying four R-3S missiles.[83]

Contrary to the Ethiopian decision to keep F-86s out of the battle, the CCS very much relied on old MiG-17s as its primary fighter-bombers. Certainly, the type was originally designed as a high-altitude bomber-interceptor. However, although excellently armed with two 23mm cannons and one 37mm for this purpose, by 1976 it was too slow and too short-ranged to match the EtAF's F-5s. The MiG-17s available to the CCS belonged to the original variant equipped with the afterburner-less VK-1 engine. As far as is known, they never received any kind of modifications making them more suitable for ground attack operations, such as carrying combined loads of fuel and bombs or unguided rockets. Whenever bombs or pods with unguided rockets were installed, drop tanks had to be left out, dramatically decreasing their range, especially under 'hot and high' conditions over Ogaden. Nevertheless, once free of

83 Kondaurov, V. N., *Runway with Length of Life* (in Russian), retrieved from 'www.testpilot.ru' in 2006 & interviews with several former CCS pilots provided on condition of anonymity. Of interest is that Kondaurov misidentified the recipient of his report as 'Ethiopia', even as of 1976 there was certainly no kind of military cooperation between Addis Ababa and Moscow. Indeed, other Russian publications reveal that the results of Konaurov's testing were revealed to Ethiopians only after the start of the Ogaden War, supposedly 'sometime in late July 1977'. Ethiopian sources deny even this, stressing that no Soviet advisers began working with the EtAF before sometime in late Sept. 1977, by which time the battle for air superiority was already decided. Even more so, interviewed EtAF pilots denied they had any kind of requirement for any kind of Soviet intelligence about comparative performances of the F-5E and the MiG-21. They stressed that their training in the USA, and especially dissimilar air combat training before the Ogaden War, proved more than sufficient to prepare them for fighting Somali MiGs.

The nemesis of Somali MiGs: although EtAF F-5Es were few in numbers, their flexible and highly intensive deployment supported by the early warning radar station at Karamara Pass, turned Ethiopian Tiger IIs into true 'force multipliers'. Although showing one of eight F-5Es never delivered to Ethiopia, this photograph is representative of the general appearance of EtAF Tiger IIs during the Ogaden War. (Tom Cooper Collection)

Insufficiently supported by early warning radars and flown by poorly-trained pilots, Somali MiG-21MFs proved no match for Ethiopian F-5Es. This example survived the war, only to be found by US troops in derelict condition at Mogadishu IAP in 1992. (Jacques Guillem Collection)

The EtAF operated only three Canberras at the start of the war, but they proved highly effective, especially during intensive interdiction operations against Somali supply lines and depots. One Canberra was lost due to technical malfunction, probably caused by ground fire. The crew ejected before reaching Debre Zeit AB. (BAe Heritage)

About twelve Mi-8Ts delivered by the USSR to Ethiopia in late 1977 entered service with 14th Helicopter Squadron EtAF and were deployed together with several examples operated by the Army Aviation. These were used intensively for heliborne advances of the 1st Paracommando Brigade, deep behind the flanks of the SNA's positions in the Harar and Jijiga area, eventually resulting in the collapse of the latter. Of interest is its 'desert style' camouflage pattern applied on this and other EtAF Mi-8s, which was unusual for deployment over Ethiopia. (Hervé Desallier via Albert Grandolini)

The F-5A serial number 661 was the first ever delivered to Ethiopia. It was retained in this livery during the Ogaden War, when it flew a number of combat sorties, often armed with LAU-61 pods for unguided 2.71in rockets. Grounded at some point during the war, possibly due to some sort of combat damage, it was subsequently used as a source of spares, and was last seen in operational condition inside one of the maintenance hangars at Debre Zeit AB in 1979.

The F-5A serial number 662 was one of the original Ethiopian Freedom Fighters, delivered in 1966, but already wearing this standardised pattern (variously named as 'Asia Minor' or 'Flogger') consisting of dark sand FS20400, dark earth FS30140 and dark green FS34079. Unusually, the anti-glare panel in front of the cockpit (in dark green) was left in place. It is shown carrying one of the heaviest payloads deployed during the war, primarily during the fighting between Harar and Dire Dawa in August and September 1977, carrying one multiple ejector rack with six Mk.82 GP-bombs.

The F-5A serial number 677 was one of the ex-Iranian Freedom Fighters delivered to Ethiopia in the 1973–1974 period. It entered service with the 5th Squadron already camouflaged in this fashion. It is shown armed with one Mk.82 GP-bomb with Mk.15 Snakeye retarding fins, while insets are showing one M-117 GP-bomb and a BLU-27 napalm tank, which were frequently deployed during the Ogaden War. This aircraft survived the war and was sold to Iran in 1990, where it is still operational and wearing the serial number 2-7254.

The F-5E serial number 418 (FY-number 74-1521) was the second Tiger II delivered to Ethiopia, and is shown armed with one AIM-9B Sidewinder missile and a BLU-27 napalm tank. The aircraft was one of four that survived the Ogaden War, during which it was used to score at least one air-to-air victory. In 1984 it was sold to Iran, where it is still operational and wearing the serial number 3-7319.

Very little is known about the EtAF F-5E with serial number 430. Belonging to the first batch of eight aircraft that were delivered in 1976, it should have worn one of the serials in the range 417–424. However, like all the other Ethiopian Tiger IIs, following delivery it received the national insignia originally worn during the times of the IEAF, and then a new serial number, probably designed in order to cover up the number of the actually delivered aircraft. This aircraft survived the Ogaden War and subsequently received six kill markings applied under the right side of the cockpit, as shown on the inset. Its subsequent fate remains unknown. While Iran is known to have purchased all four surviving Ethiopian F-5Es in 1984, it was never sighted in service in that country.

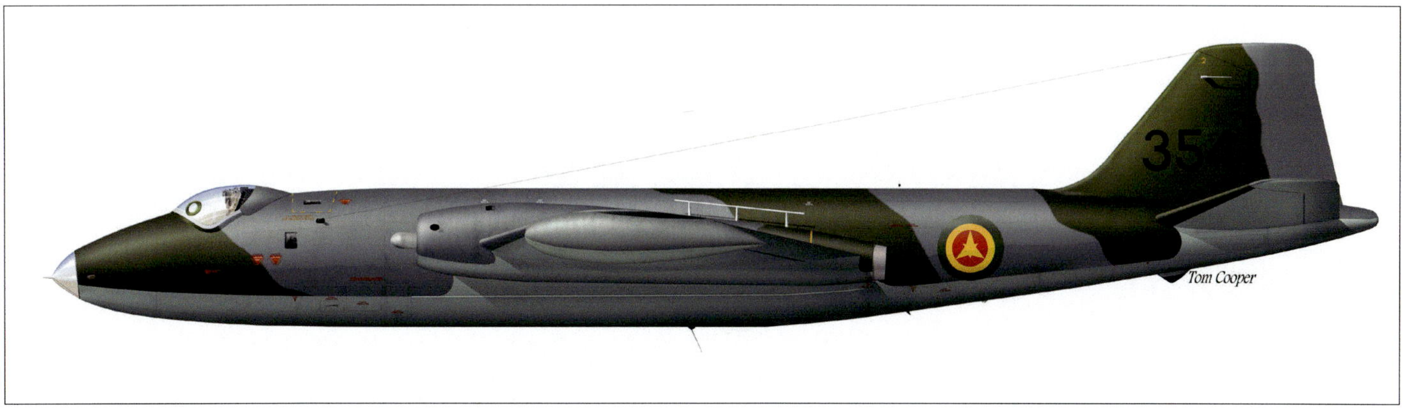

The Canberra B.Mk. 52 serial number 354 was one of two bombers of this type that survived the Ogaden War. Like all the other examples it was delivered wearing a standardised camouflage pattern consisting of dark sea grey and dark grey on top surfaces, Sky S on lower surfaces, and roundels applied in six positions. Its derelict wreckage was last sighted at Debre Zeit in 1999.

This is a reconstruction of one of thirteen MiG-17Fs delivered to Ethiopia between September and November 1977 from Yemen. The aircraft was apparently never camouflaged, and thus still showing clear traces of its former Soviet Air Force markings (including red stars on the fin and the number '02' on the forward fuselage) when sighted for the last time, at Dire Dawa, several years ago.

This was another of the ex-South Yemen Air Force MiG-17Fs delivered to Ethiopia in 1977. It appears that most of these aircraft were camouflaged in this fashion. All were flown only by Cuban pilots during the Ogaden War.

A reconstruction of the MiG-21MF serial number 1050. Probably newly-built and delivered straight from the USSR, they were not taken up by the EtAF but only flown by Cuban pilots from Dire Dawa in early 1978. What happened with these aircraft afterwards remains unknown. The aircraft is shown as armed with a ZAB-350 napalm tank.

Although designed, developed and equipped as a lightweight interceptor, as there was no threat from CCS interceptors, the EtAF MiG-21bis's were primarily deployed as fighter-bombers, usually armed with various 250kg bombs or, as shown here, with UB-16-57 pods for unguided 57mm rockets. Although theoretically standardised, their camouflage pattern – consisting of beige (BS381C/388) and green (BS381C/283) on top surfaces, and light admiralty grey (BS381C/697) on undersurfaces, differed from aircraft to aircraft.

The four MiG-21Rs that should have been delivered to Ethiopia in December 1977 (only two have been visually confirmed so far), were painted in a slightly darker fashion than the MiG-21bis's, namely in deep buff (FS30257), and a shade similar to red oxide (FS30166) on top surfaces. They were usually equipped with a 'Type D' reconnaissance pod, installed under the centreline and equipped with reconnaissance cameras, as shown here.

MiG-23BN fighter-bombers with wings of variable geometry were latecomers in the Ogaden War, but they did fly a number of combat sorties during the final days of the conflict, and additional strikes against Somali targets afterwards. They arrived painted in standard camouflage pattern for all examples of this variant exported in the 1970s, including sand (FS13523), dark earth (FS20095) and green (FS30098) on the upper surfaces and sides, and light blue-grey (FS35622) on the lower surfaces. The Soviets delivered them to Ethiopia together with a wide range of weapons, including UB-16-57 pods for unguided rockets (shown installed on the aircraft), and (from left to right) FAB-250M-54 and FAB-250M-62 GP-bombs, FAB-500M-62 GP-bombs, two variants of RBK-250 CBUs, and ZAB-350 napalm tanks.

A reconstruction of the MiG-17 CC125, as found by UN troops in Mogadishu in 1992. This aircraft was one of many CCS planes that had the winged leopard insignia applied on the front fuselage, usually in yellow and black. It was camouflaged in beige and olive green on the upper surfaces, and appears to have had its undersurfaces left in 'bare metal', or painted in light grey. Inset is showing a reconstruction of its top-side camouflage pattern.

Another MiG-17 found at Mogadishu in 1992 that was, like the example shown above, almost certainly a veteran of the Ogaden War, was serial number CC-134. The aircraft was painted in the usual camouflage pattern of beige and dark olive green. Contrary to other CCS MiG-17s it received only the national insignia on the fin. It is shown armed with a UB-16-57 pod, installed instead of the (more usual) drop tank. As delivered to Somalia, the type was unable to carry both external fuel tanks and weapons at the same time, which greatly diminished its effectiveness.

This reconstruction of the MiG-17 serial number CC-136 includes the usual arsenal of bombs delivered by the type during the Ogaden War, including (from left to right), FAB-50M-4, FAB-50M, FAB-250M-46 and the RBK-250 cluster bomb. A single FAB-100M, which was used in large numbers, is shown installed underwing.

This MiG-21MF serial number 252 was one of only a handful of Somali aircraft of this type that survived the Ogaden War. It wore the standardised camouflage pattern consisting of beige (BS381C/388) and green (BS381C/283) on top surfaces, and light admiralty grey (BS381C/697) on undersurfaces, which is shown in a worn-out condition. The vast expanses of the battlefield in Ogaden usually necessitated these aircraft to operate in this configuration, carrying three drop tanks and only two R-3S AAMs.

Like all Somali MiG-21MFs, serial number 256 was painted in a standardised camouflage pattern consisting of beige (BS381C/388) and green (BS381C/283) on top surfaces, and light admiralty grey (BS381C/697) on undersurfaces. Somali MiG-21MFs were exclusively deployed as interceptors, for which purpose they could be armed with up to four R-3S AAMs, as illustrated here.

A reconstruction of one of three MiG-21UM two-seater conversion trainers delivered to Somalia in 1974. While usually assessed as lacking forward-view mirrors installed on the top of the rear cockpit (designed to enable the instructor pilot to see ahead of the aircraft), photographs from the early 1980s show them with this installation in place. All three Somali MiG-21UMs were left in bare metal overall, and had roundels applied only on their fins.

A reconstruction of CCS MiG-15UTI CC116 found abandoned inside a hangar at Mogadishu Air Base in 1992. The aircraft was camouflaged in beige and dark olive green on the upper surfaces, but its lower surfaces were apparently left in 'bare metal', as were the original 'slipper type' drop tanks, installed directly on the lower surfaces of the wing. Roundels were applied in six positions.

Little is known about the activity of the diminutive fleet of four Mi-8T helicopters operated by the CCS during the Ogaden War. It is very likely that they were deployed for support of ground units early during the war, primarily for transport and liaison purposes. This is a reconstruction of the only example ever photographed, which appears to have been left in dark olive green overall, probably as painted before delivery. Only the bottom of the cabin was painted differently, probably in light admiralty grey. Except for its serial number, notable is the original CCS fin-flash, applied on the root of the boom, very seldom seen on its aircraft in the 1970s.

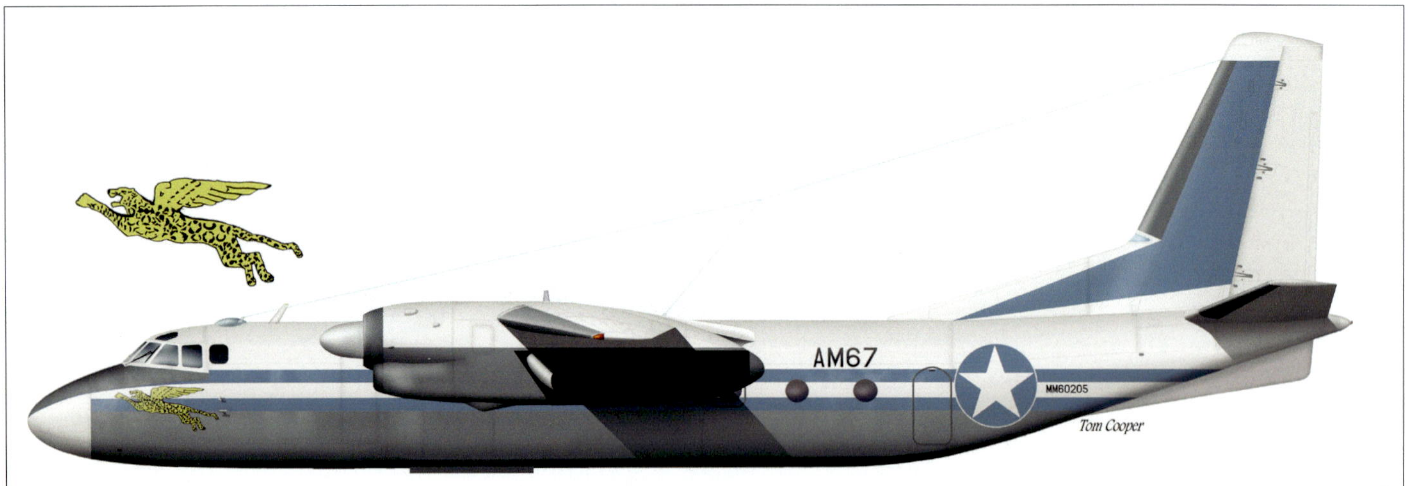

Early during the Ogaden War, Somali An-24s were often deployed to haul supplies for the SNA to Jijiga airfield. In late 1977 and early 1978, they were also deployed as bombers, almost exclusively by night. All were painted in this fashion, and it appears that most received the winged leopard insignia on the front fuselage.

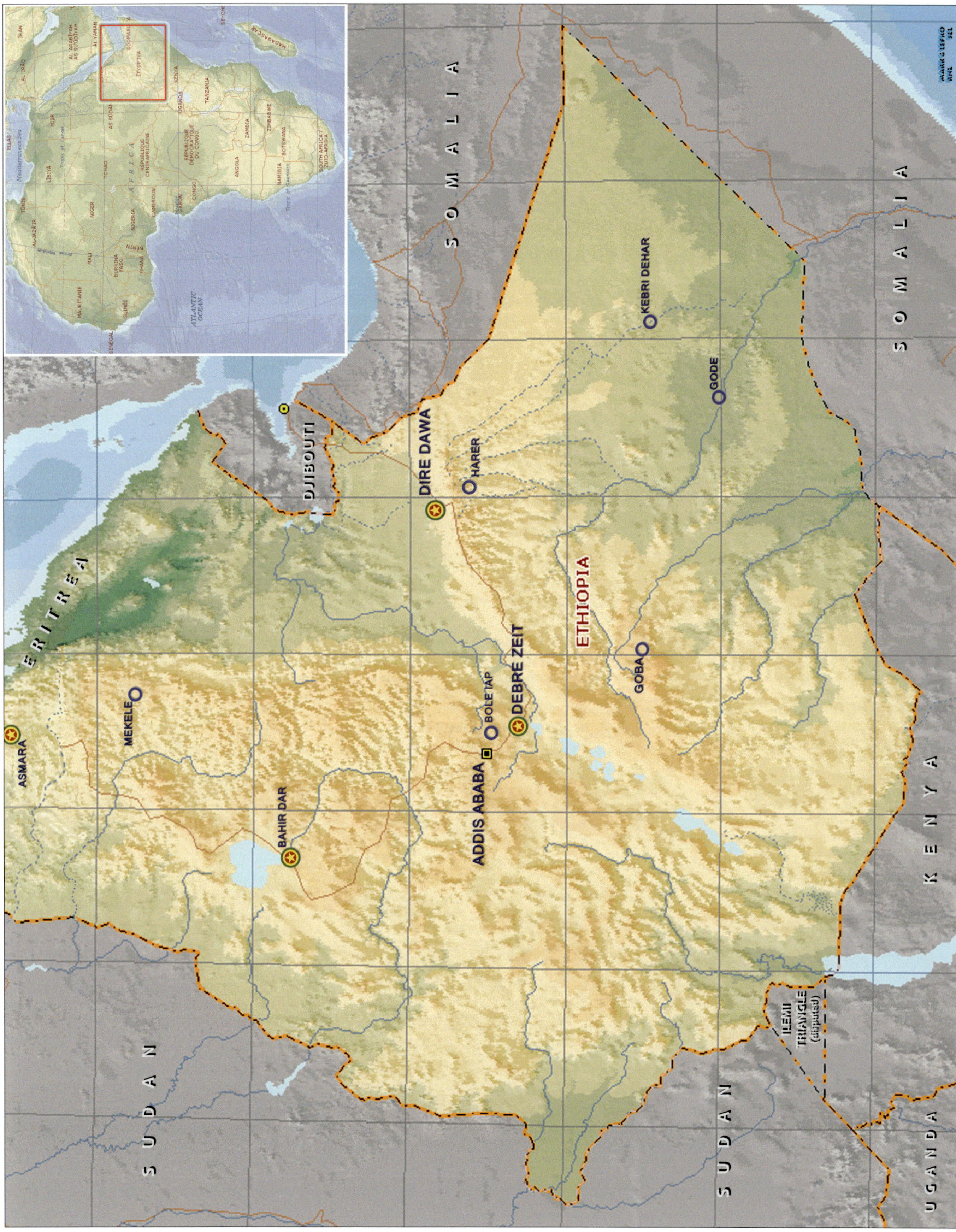

Map of Ethiopia as of 1977, including all air bases and the most important airfields.

Map of Somalia, with air bases and the most important airfields.

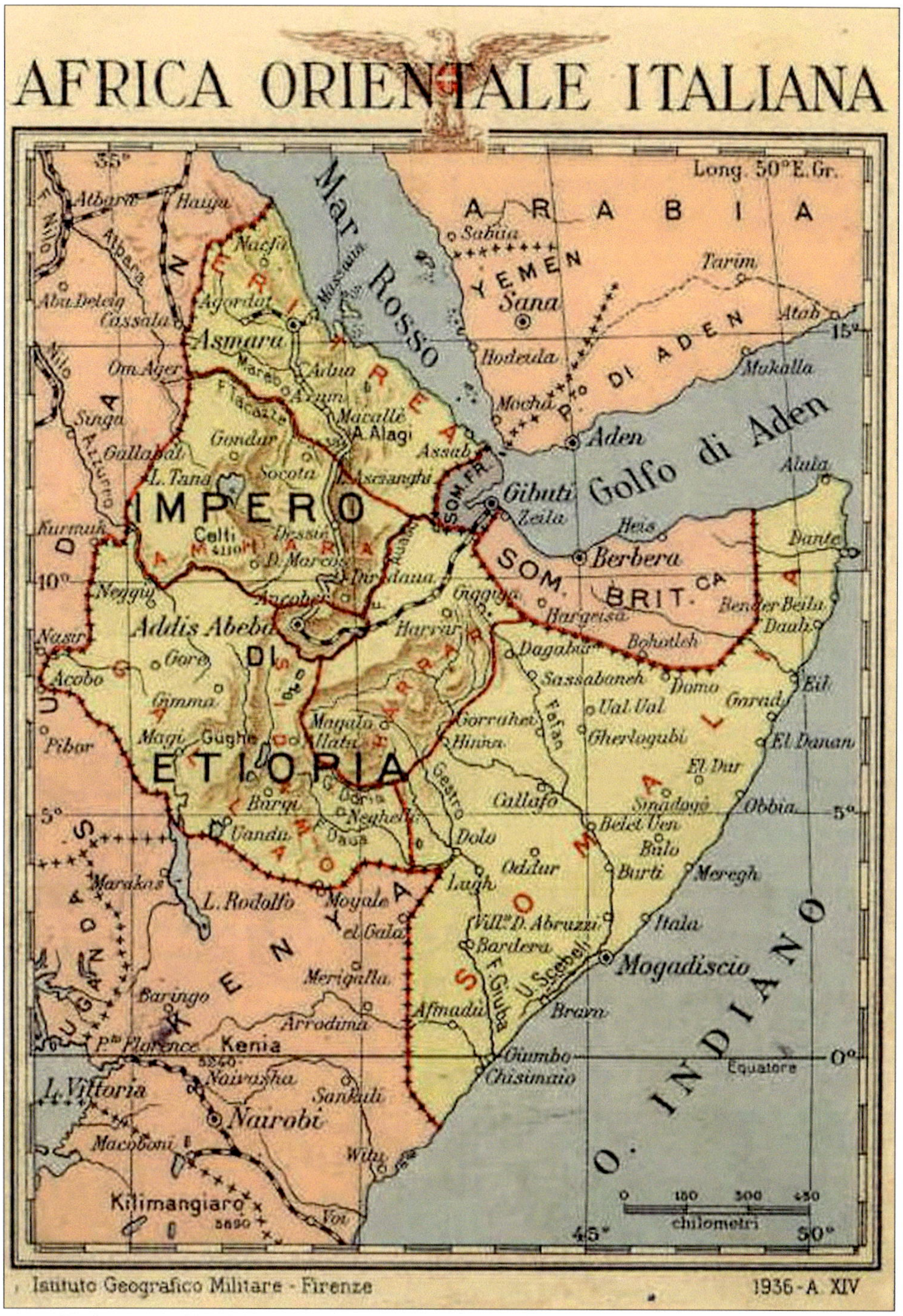

Map of 'Italian East Africa' from 1936. (Istituto Geografico Militare)

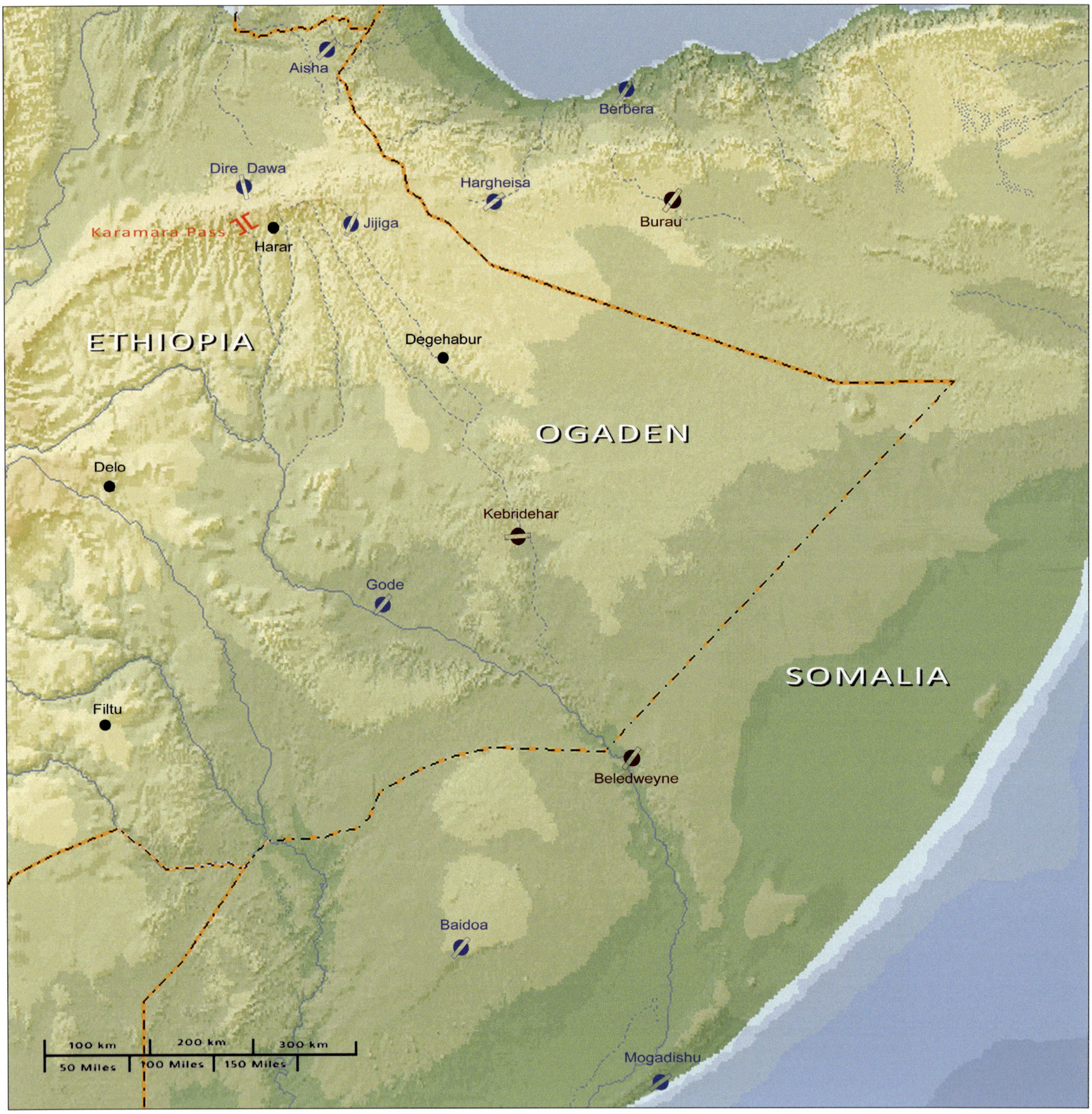

Map of Ogaden with the most important air bases and airfields in the area, and in neighbouring Somalia.

Local farmers picking water from the Shabele River in the Gode area.

Nomadic herders – primarily ethnic Somalis in the Harar area; they dominate the population of Ogaden. (US DoD)

Although much of Ogaden is characterised by a semi-desert of sandstone and limestone, the Harar Plateau in the north-east is predominantly green and dotted by farms. (US DoD)

Tewodros II (known as Theodore in the West) is considered the founder of modern-day Ethiopia. (National Museum of Ethiopia)

Two IEAF F-86Fs at Kamina AB in southern Congo. Of interest are the double sets of drop tanks necessary for long flights between different air bases in that huge country. (Harry Nanneson via S. N.)

Pilots of 1st Fighter Intercept Squadron IEAF posing in front of an F-86F. Standing from left: Bezabih Petros, Alemayehu Gondere, Taweke Assefa, Ashagre Mekonnen, Neguisse Zergaw, Habtewold Gebrewold, Asemare Getahun, and Techane Mesfin (CO 1st Squadron). Kneeling from left: unidentified and Goitom Asfeha. (EtAF via S.N.)

Sadly, photographs of intact Somali MiG-17s and MiG-21s from the 1960s and 1970s respectively, remain unavailable. This shot was taken by US troops when they occupied Mogadishu IAP in 1992, and shows one of three MiG-21UMs delivered to the CCS in 1974. (Tom Cooper Collection)

Cuban advisers seen training recruits of the People's Militia at Camp Tatek in the summer of 1977. Although already present in Ethiopia by that time, Cubans were still months away from becoming involved in fighting on the battlefields of the Ogaden War. (Albert Grandolini Collection)

Col Abdullah Askar was in command of Somali Army units that attacked Gode. For his success, and the supposed destruction of 'two Ethiopian divisions', he earned himself the nickname the 'Lion of Gode'. (via Sheikh Hassan)

CCS early warning radar site outside Berbera, as seen in autumn 1982. It consisted of two truck-mounted, mobile P-12 radars (both seen in folded position, in the background left and right) and one P-15 radar (foreground, in erected position). (US DoD)

An UH-1H Huey helicopter of the Ethiopian Army Aviation preparing for takeoff from Debre Zeit AB, right next to one of the EtAF's Mi-8Ts. Although fiercely anti-US, even General Petrov learned to appreciate the more comfortable Huey. (Robert Szombati Collection)

The MiG-23BNs arrived over the battlefield only during the last weeks of the Ogaden War, but saw much action over this area during subsequent months. It evolved into the major fighter-bomber of the EtAF during the 1980s and long after. (Herve Desallier via Albert Grandolini)

Although rebuilt through deliveries of Chinese- and Italian-made aircraft and helicopters after the Ogaden War, the CCS never recovered from the losses sustained during that conflict. Ultimately, it was destroyed in the course of a civil war that ruined Somalia in the late 1980s and early 1990s. Here, a row of abandoned MiG-21s as found by US troops at Mogadishu. (Jacques Guillem Collection)

Following his return from captivity in Somalia in 1988, Lagesse Teferra was advanced in rank to Major-General and highly decorated (between others also by Raul Castro), as can be seen in this photograph. (via S.N.)

While in the process of being replaced by MiG-21s, the F-86Fs continued serving with the EtAF for a while after the Ogaden War. Thus, when a transport aircraft carrying a delegation of the Hungarian Air Force arrived over Ethiopia in 1979, it was escorted by these two Sabres. (Robert Szombati Collection)

Pre-war photograph of a CCS Mi-4 helicopter, seen carrying a big Somali flag with a portrait of Siad Barre. This type was operated, together with some four Mi-8s, by the sole Helicopter Squadron. Precise details about participation of this unit in the Ogaden War remain unknown.(Albert Grandolini Collection)

bombs or rockets, MiG-17s were very manoeuvrable and capable of performing highly complex aerobatics, although lacking engine thrust for sustained turning performance. The Somalis and their Soviet advisers assessed them as superior even to F-5Es. Combined with the expectation of meeting only little or no resistance from the Ethiopian airmen, they were certain that they could survive over the battlefield. Worse still, what they did not know was that EtAF pilots had received very intensive and realistic air combat manoeuvring training both in the USA and at home, and that the F-5's weapons were to prove highly reliable and effective. It was only after the war that the CCS pilots realised that they had flown much too little before the war and received next to no training in air combat.[84]

84 Interviews with former CCS MiG-21-pilots, provided on condition of anonymity, 2003 & Ashenafi, interview, 2001.

If not lack of flying, the poor training of their pilots should have become obvious to top CCS commanders when the air force experienced two unpleasant incidents in the months before the war. On 25 March 1977, two MiG-17s intercepted and attacked French Naval Aviation Lockheed P-2H (P2V-7) Neptune reconnaissance aircraft over the Gulf of Aden, monitoring CCS activity. The French pilot initiated a series of evasive manoeuvres at very low level and avoided several strafing runs before the MiGs were so short of fuel, they had to return to Berbera. Just a few weeks later, one CCS MiG-17 is known to have crashed, killing its pilot, in the course of a landing attempt at Berbera after a similar action that resulted in the forcing down of an Air France Boeing 707. [85]

Its aircraft and pilot training were not the only problems facing the Somali Air Force. Another major issue was the lack of an effective radar network that could support operations over Ogaden. The CCS operated SA-2 and SA-3 missiles that were supported by powerful early warning radars, but these were deployed solely for defence of major air bases inside Somalia, and could thus provide next to no meaningful coverage of the airspace above the battlefield. Furthermore, they were hardly ever even activated, as recalled by Koshelev:

> Only one SA-2 site was in a position to defend our air base, but its radars were seldom turned on and then only in order to check their function and keep the crew current. What the Somalis did in terms of preparing their air defences was not even distantly similar to combat conditions.

As the entire SNDF was soon to find out, its and CCS' preparations for air defence were entirely insufficient.

85 Jacques Guillem, former Air France B707 pilot, interview, June 2009.

CHAPTER 4
INVASION

Battle of Gode

At 0300hrs on 13 July 1977, two SNA armoured and three mechanised brigades under the command of Col Abdullah Askar launched their attack into south-eastern Ogaden, rolling in the direction of Gode. A few hours later, at dawn, the CCS fighters and bombers flew a series of 50 combat sorties against Ethiopian units in the Gode area. Although defending successfully, over the next two days the 5th Ethiopian Brigade deployed to protect the town, was to suffer heavily from air raids and Somali artillery.[86] The fighting continued for a few days longer and it was here that the EtAF suffered its first loss of the war. On 14 July 1977, a formation of three F-5As returned to Dire Dawa AB when the CO 24th Brigade called with a request for close-air-support (CAS). The leader of the F-5-formation, Berhanu Kebede, had problems with the radio in his plane, but there was no time for technicians to take care of it. He thus assigned his wingman, Asmare Getahun, as formation leader, while Lule Gebre Median joined them as number three. Unable to communicate with other formation members by any other means than hand signals, and with Gode airfield out of use due to heavy shelling by Somali artillery, Berhanu followed his formation into attack. The Somalis fired at least one SA-7 at his aircraft, and this obtained a steady lock-on. The other pilots saw the missile but Berhanu could not hear their calls and the MANPAD flew right into the left engine of his F-5A before coming out the side of the aircraft. Although it failed to detonate, the SA-7 caused enough damage to force Berhanu Kebede to eject south of the town. Fortunately for him, the main area of the fighting was further south and the wind coming from that direction blew him closer to the local Ethiopian base, which he reached after walking for several hours. Later on, Berhanu was picked up by an army helicopter and brought back to Dire Dawa AB.[87]

From that time onwards, Ethiopian pilots, still not knowing they were facing the firepower of the SNDF, but believing they were fighting the WSLF, were forced to adapt their attack profiles in order to avoid unnecessary exposure to SA-7s. The profile in question included a dive-bombing pass from medium altitude with engines on idle and then engaging afterburners to swiftly climb outside the SA-7's engagement envelope. Other bomb and rocket-delivery profiles included approaching from very low altitude and turning directly into the sun upon weapons release, thus presenting the

86 Tareke, p. 644 & Cooper et al., *African MiGs Vol. 2*, pp. 101–102.

87 Berhanu, interview, Oct. 2008.

The DC-3 was still the backbone of EAL's airliner fleet as of 1977. One of such aircraft was destroyed by Somali MiGs during an attack on Harar airfield on 21 July 1977. (Jacques Guillem Collection)

Somali Army T-34/85 tank and its crew shortly before the Ogaden War. (Albert Grandolini Collection)

Early during the war the SNA and WSLF captured a large number of Ethiopian soldiers. (Albert Grandolini Collection)

missile with a 'decoy' of sorts.

Gode AB and the nearby base of the 5th Brigade were soon battered by Somali bombardments. Although reinforced by the 79th Militia Brigade, and ably supported by the EtAF, the attacks of which caused heavy losses to the Somali infantry, the Ethiopians could not hold their positions and the town fell to Col Askar's troops on the morning of 25 July. The 79th Militia Brigade was virtually wiped out in the process of withdrawal, suffering a loss of over 1,900 killed in action (KIA) and wounded in action (WIA), while the 5th Brigade came away in relatively good condition, but not without losing 79 KIA and 8 WIA.[88]

Rapidly rolling in a south-westerly direction, foremost by night in order to avoid detection, the northern group of Somali forces first reached Aisha on the evening of 16 July. Defended by the 24th Brigade EA and 752nd Militia Battalion, the town was assaulted at 4:30am the following morning, with support of CCS MiG-17s and Il-28s and top cover from MiG-21s, followed by an extensive artillery bombardment. The sole Il-28 involved actually dropped its bombs from a very high altitude, so that these hit the area ahead of Ethiopian front lines and caused no damage. The Ethiopians, who later recalled the bomber flying like a civilian airliner vectored by ground control to make an approach for landing, reacted by scrambling a pair of F-5Es from Dire Dawa and also several ground-to-air attacks, but could not provide sufficient help.

The condition of Ethiopian ground forces in the Harar area was soon critical as well. After failing to stop the Somalis at Kebridehar and then refusing to take up new positions, the 9th Brigade was forced to retreat into that town in state of disarray. On 21 July, the CCS MiGs hit the Harar airfield and destroyed an Ethiopian Airlines (EAL) DC-3 passenger aircraft on the ground.[89] The 11th Brigade refused to give up, enduring withering artillery fire and dozens of air strikes until the end of July, when it was ordered to withdraw to Jijiga. By 8 August, the Somalis captured Delo, Filtu and Elkere, and their columns were well underway in the direction of Dire Dawa, Harar and Jijiga.

88 Ibid., & Tareke, p. 644.

89 Registration of the aircraft in question was ET-AAP. While all the available Ethiopian and Somali sources stress the EAL DC-3 was destroyed on the ground, Babich and Kotlobovskiy (*MiG-21 in Local Conflicts*, p. 31) credit it to a CCS MiG-21MF and claim it was the first air-to-air victory of this war, and that the crew of two was killed.

M113s of the Ethiopian Army captured by Somalis at Gode. (Albert Grandolini Collection)

One of the C-47s operated by the 21st Transport Squadron EtAF which was shot down by two Somali MiG-17s on 21 July 1977. This was the first confirmed air-to-air victory of the war. (Dave Becker)

Except C-47s, the EtAF also deployed its C-119Ks to drop supplies to isolated garrisons. The type of aircraft was also used as an airborne command post and radio-relay 'station'. (Dave Becker)

Bezabih Petros scored the first officially confirmed MiG-kill of the Ogaden War. (via S.N.)

The first Kill

Because Addis Ababa and Mogadishu severed their diplomatic relations in early 1977, and because Somalia never officially declared war on Ethiopia, the Derg initially failed to comprehend the situation and their first reactions were rather slow. Mobilisation of the Ethiopian armed forces was announced on 15 July, but it took the military several weeks to accept over 250,000 new recruits and reservists for which it lacked even uniforms, not to mention arms.

On the contrary, euphoric over initial success, the government in Mogadishu began issuing fantastic claims, reporting the destruction of at least eight EtAF aircraft in attacks on different airfields during the first week of war, all on the ground. In fact, while the SNA continued its advance, the air war began developing into anything but a positive manner for the CCS. Namely, while involved in CAS operations for advancing ground forces, the CCS fighter-bombers were exposing themselves to interception by F-5Es of 9th Squadron, skilfully directed by the TPS-43D radar positioned at Karamara Pass, and soon began suffering losses in air combats.

Although the first two air battles are known to have occurred already on 16 and 20 July, the EtAF did not credit any of its pilots with a confirmed win. One of the reasons was the intensity of operations in this chaotic period (most of the pilots flew an average of more than three combat sorties a day), and the other was very stringent rules for officially crediting any 'kills'. Namely, aside from usual methods of confirmation through inspection of wreckage and gun-camera film, or confirmation by wingman, the EtAF expected its pilots to score hits on enemy aircraft with their weapons in order to be credited with a win. Therefore, no kills were credited for so-called 'manoeuvring kills', i.e. when Somali losses were caused by a pilot's mistake that led to a crash. Lacking recent flying experience, a number of CCS pilots crashed their aircraft while engaged in air combat with EtAF fighters, especially early during the war, and thus a significant number of theoretical victories were never credited to any Ethiopian pilots. In other cases, Ethiopians intimidated their opponents to a degree where several cases were reported of CCS pilots ejecting even before fired upon.[90]

On 21 July, the EtAF deployed most of its transport aircraft to launch a re-supply effort for a number of army garrisons cut off by the Somali advance. Still unaware of the Somali invasion, the air force did not provide C-119s and C-47s with any kind of top cover. On one occasion the transport aircraft flown by Maj Worku Woldemariam, with Ashagare Mekonnen as co-pilot, Master Technician Abebe Melku and Senior Technician Dendana Redat as loadmasters, was carrying a load of ammunition and other supplies to Geladin when they were intercepted by two MiG-17s. The Somalis attacked the lumbering transport aircraft and Worku attempted to evade their fire by flying tight turns at low level, but the C-47 was eventually hit in the left wing. Dendana then notified the pilot that the plane was on fire and they made a safe belly landing in the desert. By the time the plane came to a stop, Dendana and one of two army soldiers who were on board had been killed by shells from Somali MiGs. The five survivors abandoned the crash site and attempted to escape, but all were captured by Somali troops the next day.[91]

It was only after Worku's C-47 failed to return to Debre Zeit that the EtAF began flying combat air patrols over Ogaden. A rule was

90 Ashenafi, interview, 2001 & Girma, p. 55.

91 There is some uncertainty over the exact date of this mission. According to other Ethiopian sources it occurred on 24 July. Kotlobovskiy dated it as 21 July, as an EAL DC-3 was shot down in Jijiga by a CCS MiG-21, which indicates he had also misinterpreted this EtAF C-47 for the EAL DC-3 destroyed on the ground in Aisha, like Babich. On the contrary, survivors of the crew (interviewed on condition of anonymity), stressed they came under attack of two MiG-17s.

Lagesse Tefera, the most successful EtAF F-5E-pilot of the Ogaden War. (via S.N.)

Lagesse Teferra as seen in the USA, during his conversion training on F-5Es. (EtAF via S.N.)

established whereby F-5Es would be airborne whenever transport aircraft were flying over the combat zone. Late in the afternoon on 24 July 1977, a pair of F-5Es flown by Mengistu Kassay and Bezabih Petros was vectored to intercept what turned out to be two CCS MiG-21MFs that seemingly attempted to intercept one transport aircraft. Expertly guided by the controller at Karamara Pass, the two Ethiopian interceptors rolled out directly behind their opponents. Mengistu engaged the Somali leader without success and the MiG managed to flee in the direction of Hargheisa, but Bezabih swiftly shot down the wingman.[92]

Bezabih Petros thus scored a truly 'historic' victory: not only was this the first ever officially confirmed 'kill' in the history of the Ethiopian Air Force, but also the first ever encounter between these two classic lightweight fighter designs, operated by so many air forces around the world. It was also a victory of a US-trained pilot, flying a US-made aircraft, over a Somali pilot trained in the USSR, flying a Soviet-made aircraft.[93]

9th Squadron contra CCS

Lagesse Teferra followed hard on Bezabih's heels by managing a special feat while leading a trio of F-5Es, the so-called 'Golden Triangle Formation', with Bacha Hunde and Afework Kidanu as wingmen, on 25 July 1977. While approaching their opponents, that were flying in the direction of Gode, the F-5s entered a shallow climb while gradually spreading their formation. In this fashion they entered the combat zone at high altitude while being able to protect each other, and also visually scan a large area of airspace around them. Eventually, the Ethiopians found themselves high above three MiGs and dived to attack.

According to Somali sources, the MiG-21MF formation consisted of four aircraft led by the CO Hargheisa AB, Col Mussa, with a Major and CO of the MiG-21-squadron based there as wingman. When the Somalis saw Lagesse's F-5 above them, Musa and his wingman separated, one turning left and the other right. Lagesse decided not to follow either but pulled up and accelerated back to altitude, watching behind his aircraft as the drama unfolded below him. The two Somalis completed their 180 degree turns and then slammed into each other! Both MiGs crashed in flames, killing their pilots as a result. Shocked by this sight, but undeterred, Lagesse then attacked a third MiG and shot it down with 20mm cannon fire. Meanwhile, Bacha and Afework chased the fourth MiG-21, cornered it and forced it into a series of tight manoeuvres. Lacking speed and down at low level, the Somali then lost control of his plane and crashed. Bacha and Afework overflew the crash site of the Somali fighter to verify its early demise. The battle was not yet over. Lagesse sighted and engaged a formation of four MiG-17s, shooting down two of them in quick succession using AIM-9

92 Girma, *Let History Speak for Itself*. The author of this book, Col Girma Gorfu, used to serve in the EtAF administration until 1991. His book includes a number of Amharic translations of Somali military telegrams intercepted by Ethiopian military intelligence. The authenticity of these translations was later confirmed to author by several Somali sources. In this specific case, the loss of a CCS MiG-21MF on the afternoon of 24 July was confirmed in telegram NAV /07/24/7/77/1 from the HQ of the 26th Division to 18th Brigade SNA, wired on the same afternoon, with following content: '24 1700 very urgent From 26th Division To 18th Brigade Gabeleb NAV /07/24/7/77/1 This is to notify you one aircraft assigned to 26th command has been lost at 8pm tonight. 24th July, 1977 time 17:56. Babich also mentions the loss of a Somali MiG-21MF on 24th July.

93 Although considered 'hot' at the time of the Cold War, because the F-5 and MiG-21 formed the backbone of many of NATO and Warsaw Pact air forces respectively, their clashes during the Ogaden War remained largely unknown outside Ethiopia and Somalia. Indeed, even 30 years later many foreign observers remain convinced that the F-5 and the MiG-21 never met in air combat, whether during the Ethiopia–Somalia War of 1977–1978, or the subsequent war between Iran and Iraq.

Bacha Hunde scored his only official kill of the Ogaden War on 29 July 1977. (EtAF via S.N.)

Pre-delivery photograph of the F-5E Tiger II serial number 430. Although this plane never reached Ethiopia, one of six F-5Es that did received the same serial and became the most successful Tiger II in EtAF service; different pilots scored six kills flying this aircraft during the Ogaden War. (Albert Grandolini Collection)

Although capable of reaching speeds of Mach 2 and armed with four R-3S air-to-air missiles (compared to only two that could be carried by EtAF Tiger IIs), Somali MiG-21MFs consistently proved no match in air combats with Ethiopian F-5Es. (Claudio Toselli Collection)

Sidewinder missiles. The last two MiG-17s managed to get away, primarily because the Ethiopian formation leader was now out of missiles and short on fuel.[94]

Although it was the manoeuvring of Ethiopian fighters that caused three of these six kills, and thus only three of Lagesse's claims were officially credited by the EtAF, the loss of four precious MiG-21MFs and two MiG-17s together with six experienced pilots was a terrible blow for the CCS. By that time, the Somalis are known to have lost eight MiG-21s and five MiG-17s in the war, most of these in air combats, with only a single kill against a C-47 to show in return.[95] In an attempt to provoke the EtAF into air combats and extract revenge, the CCS then launched a number of raids against Ethiopian Army positions during the following days. None of these were successful however, as the MiGs lacked the range and weapons suitable for causing sufficient damage. On the contrary, the Somalis only exposed themselves to interceptions by Ethiopian F-5Es and suffered additional losses.

On 26 July, Lagesse Teferra and Bezabih Petros were flying escort for a C-47 on a supply drop mission when the radar at Karamara advised them about a pair of MiG-21s that was approaching Dire Dawa over Chinahaksen. This time, the Ethiopians failed to sneak up on their opponents undetected, but as the MiGs parted on sighting them and attempted to sandwich the F-5s, both Ethiopians turned after the Somali leader and swiftly manoeuvred themselves into a favourable position. The MiG did several hard turns in an attempt to shake them off, but it was too late; the Sidewinder fired by Petros detonated near the fin. The Somali fighter remained airborne but hard turns decreased its speed, enabling Lagesse to approach and open fire with 20mm cannons. Spewing pieces of debris, the MiG-21MF fell into a final spin towards the ground. The other MiG disappeared.

Apparently on the same day, the CCS also lost a MiG-17 in the area of Karamara Pass. The plane was involved in a CAS sortie when it received a direct hit from a 40mm Bofors L60 flak that caused it to enter a right turn and slam into Mount Gurach. An Ethiopian Army patrol sent to check the crash site found the body of a dead Somali pilot with the name Hussein on his flight suit.[96]

During the following days, pilots of 9th Squadron EtAF refined their tactics and began setting ambushes for Somali MiGs. One frequently employed tactic was to send a pair of F-5As on a CAS sortie and let them climb to a flight level of 9,144m (30,000ft) so they would be certain to be acquired by Somali early warning radars, prompting the CCS to scramble a pair of MiG-21MFs from Hargheisa. Closely monitoring Somali radio communications and tracking the MiGs with the help of the TPS-43 radar at Karamara, the Ethiopians would then scramble a pair of F-5Es from Dire Dawa and send them to intercept. By remaining at low altitude, and in strict radio silence, the Tigers could approach undetected by the Somalis. Alternatively, when operating over areas well outside the

94 Former CCS MiG-21 and F-6 pilots were interviewed on condition of anonymity. Sadly, the source in question remains reluctant to reveal full names of the Somali pilots involved. Furthermore, although the Ethiopians were certain to have engaged only three MiG-21s (and four MiG-17s), their recollections, recorded at different places and times, do confirm the presence of four MiG-21s. Interestingly, Kotlobovskiy also mentions four MiG-21s, although dating this clash as 26 July 1977.

95 Kotlobovskiy (p. 31) states that the Ethiopians reported nine kills against the Somalis by 27 July 1977, of which eight were MiG-21s, but also that Somali MiG-21s destroyed another Ethiopian DC-3 while attacking Jijiga airfield, on 21 July. Babich specifies Somali losses as thirteen MiG-17s and twelve MiG-21s without a single EtAF loss in return, all before 1 Sept. 1977. Contemporary Ethiopian media claimed up to 23 CCS aircraft as shot down, by all means, during this period, including nine or ten in air combat. This is one of a few cases where Ethiopian and Russian sources are at least in rough agreement with each other.

96 Both of these engagements were witnessed by several Ethiopian Army units stationed in the area, and their survivors provided recollections on condition of anonymity.

Still from a video showing the EtAF F-5A serial number 676 at Dire Dawa, early on during the Ogaden War. Notable is the heavy wear of the camouflage pattern, but also wing-tip fuel tanks left in their original ('Aluminium-Silver') colour. (via Pit Weinert)

Somali radar coverage, F-5Es would climb before attacking. As soon as they were close to the MiGs, ground control would advise them to activate their radars, climb and attack from bellow, above or behind, still entirely unseen by their opponents. It was in this fashion that Bacha Hunde scored his first and only confirmed victory of the war, again in air combat with a CCS MiG-21MFs east of Dire Dawa, on 29 July 1977.[97]

Hargheisa Raid

The EtAF did not limit its operations to those of air defence and the resupply of ground troops. On the contrary, the F-5s, that continued deploying to Dire Dawa AB in the morning and withdrawing in the evening, flew more than 300 close-air-support and interdiction sorties during the first two weeks of war. With SNA units approaching Dire Dawa, Tigers and Freedom Fighters flew over ever shorter ranges, eventually requiring no carriage of drop tanks. All four underwing and one centreline hardpoints could be loaded with bombs, mostly US-made Mk.82s and M-117s, launchers for unguided 2.71in (68mm) calibre rockets and F-5Es could still carry Sidewinder AAMs on their wingtips. Interdiction sorties, aimed at long supply columns of the Somali National Army, proved highly effective. These were fully exposed in the semi-desert, and suffered extensive losses to air strikes. Canberras added to the carnage, flying sorties from Debre Zeit and Dire Dawa. Reconnaissance sorties undertaken by the sole RF-5A revealed that time and again up to 20 Somali vehicles were destroyed in a single pass by two bomb-armed aircraft. Because of such operations, generally hampered only by frequent thunderstorms that occur over Ogaden during the summer season, it frequently meant Somali armour units had to attack built-up areas without full ammunition supply or without the necessary infantry and artillery support. The Ethiopian militia, armed with nothing more than rifles and Molotov cocktails, was therefore often free to assault them from the flanks and cause losses.[98]

At least as important were several attacks flown by Canberras and F-5As against targets in the Hargheisa area, in early August 1977, which culminated in a major raid by four F-5Es against the local air base, where Ethiopians caught the CCS with its 'underwear down', as the pilots called it. Flown on 12 August 1977, this raid not only took the Somalis completely by surprise, but also destroyed several aircraft on the ground, as recalled by Koshelev:

> Two MiG-21s just took off and I was in the tower, from where I had a good look at the planes as they disappeared in the distance. Tractors then towed several Il-28s to the runway. At that time no spares could be ordered from the Soviet Union and had to be purchased from China, where this type was still in production. That's why Somalis were trying to turn on the engines only in the last minute before takeoff. While watching this procession, I observed two dots in the distance, towards the West. At a second look I was surprised to see four aircraft approaching at low altitude. I asked the flight controller what was coming our way. Without any rush, without watching out of the window and while drinking his coffee, he answered that our aircraft are returning. 'How the hell ours?', I thought; they had started out only a few minutes ago and if nothing had happened they would report on radio before returning ... Within seconds, the four 'dots' took the form of four F-5s and in the next moment a series of heavy bombs hit the runway. The second pair then hit one of the Il-28s with unguided rockets. The first pair came back to hit the fuel and lubricants depot, with the second pair coming back to attack in our direction, obviously attempting to hit the control tower. That's when I recalled a chat with Volodka Nenatchov, who suggested that in war you should hit the nearest enemy airfield at the first opportunity and destroy any aircraft you can find there. Watching the two fighters turn in our direction, it was as if I heard these words again. 'Fast, get down!' We ran, or rather, we jumped down the stairs. The control tower was unusually spacious and that's why all the personnel kind of got lost in all that space, but there were lots of people on the stairs. Nobody observed rank or nationality, it was only important to get out of there. Luckily for us, the exit was nearby. While running out we could hear the shells hitting the roof and the wall over us. The aircraft thundered low over our heads. We were lucky that they only limited their attack on strafing the control tower and they had spent their rockets on the aircraft. I took a look in the direction of our aircraft; a Somali An-26 was slowly turning to side and going up in flames. I concluded that the other Il-28s and MiG-21s must have been destroyed as well. Everything was full of smoke and I saw flashes from several detonations. Eventually, we could not find out exactly how much damage the attack had caused, because immediately afterwards all the Soviet military personnel and civilians were put under house arrest in one of the local hotels. We sat there in our rooms for over a week, with an armed guard in front of our doors. Nevertheless, we were treated kindly. Later on, they evacuated us on Soviet An-12 transports back to the USSR.[99]

Showdown in Dire Dawa

Eventually, all of these efforts and successes of the Ethiopian Air Force were in vain. By 15 August, the Somalis were in control of nearly 80% of Ogaden, and the Derg were forced to withdraw and reorganise their remaining army units. Nevertheless, while recognising that the situation in Eritrea was even more critical, for without access to the port of Asseb Ethiopia would be a landlocked state, the Derg rushed two fresh militia divisions, the 2nd and the 5th, under the command of Col Haile Mariam Aberra to northern Ogaden on 28 July, where they were to bolster the defences of Harar

97 Ashenafi, interview, 2001 & Girma, p. 55.

98 Babich, *With Foreigners against Our Own*.

99 Ibid. Note that Kochelev recalled his experience as an attack on Berbera AB. In fact, the Ethiopians did not raid this airfield until late December 1977. This would also explain why Koshelev recalled the CCS preparing 'several' Il-28s for a sortie: by the time the EtAF raided Berbera, the Somalis were left with only one bomber of that type in operational condition.

One of a few EtAF T-28Ds that were still operational as of August 1977, was destroyed on the ground during the Somali assault on Dire Dawa AB. (EtAF via S.N.)

and Jijiga.[100]

Due to the Somali advance, the EtAF was eventually forced to evacuate most of its assets from Dire Dawa AB. Although T-28s of the 3rd COIN Squadron remained in place and continued flying CAS and armed reconnaissance sorties, the F-5 were henceforth to land there only to refuel before returning to Debre Zeit AB. While on the ground, their pilots and ground crews remained on permanent standby, with aircraft hooked up to auxiliary power units and ready to scramble at a moment's notice. They thus continued striking the enemy, even under most adverse conditions, causing much damage as confirmed by intercepted Somali communications.[101]

Despite numerous setbacks, the CCS remained active. On 11 August, two MiG-21s attempting to bomb the unpaved airstrip near Aisha were both shot down under as of yet unknown circumstances. Furthermore, recognising the importance of the TPS-43D radar at Karamara Pass, Somalis repeatedly attempted to hit this installation. Both the MiG-17s sent on a mission on 14 August flew into low cloud in between the mountains and crashed, killing their pilots.[102]

Nevertheless, the initiative on the battlefield remained clearly in Somali hands and in mid-August 1977, the SNA was finally ready to launch its assault on Dire Dawa. The attack began at 0430hrs on 17 August 1977, with two brigades (including the 15th Infantry), reinforced by one tank battalion from the 14th Armoured Brigade and two artillery battalions, approaching the town that was defended by elements of the 2nd Militia Division, 201st Nebelbal Battalion, 781st Battalion (78th Army Brigade), four companies of mechanised infantry and a platoon with only two M47 tanks from the 80th Tank Battalion. Expecting the Somalis to attack Jijiga instead, the Ethiopians were caught by surprise. The 781st Battalion withdrew from the Shinile hilltop overlooking Dire Dawa and moved towards the air base. EtAF fighter-bombers reacted immediately with several attack sorties, and continued bombing the enemy in the afternoon as well, but the Somalis kept on pushing. It was during one related mission that Bezabih Petros scored his second confirmed kill of that war, by shooting down a CCS MiG-21MF east of Dire Dawa. However, by 1500hrs, the Somalis closed in and began shelling the air base, eventually enabling their tanks to break through and occupy the crucial installation, ravaging it in the process. They destroyed not only one of a few T-28Ds that were still operational, but also eight stored SAAB B.17s, the local air traffic control, fuel tanks and a nearby gas station, most of the aircraft shelters, and even the cement, cotton and meat factories nearby.

This Somali T-54 was knocked out during the battle of Dire Dawa. (Albert Grandolini Collection)

100 Tareke, p. 645; the reinforced 2nd Militia Division was assigned the crucially important area between Harar and Awash, Dire Dawa and the task of securing the railway section between Aisha and Awash.

101 Girma in *Let History Speak for Itself* provided telegrams 'TA 45/25/77', '2/1', '141405', '270215', '241 500', and '071 000', all from 14 Aug. 1977, as evidence of the EtAF's effectiveness. Especially interesting were the following three: '2/1 Very Urgent From Jemal FF17 To Guled Hargheisa N L Unless you provide aerial cover, we won't survive. We haven't heard ours but their [meaning EtAF fighter-bombers] are all over the sky.'; '270215 Very Urgent From 110 To 26th Command Hargheisa Enemy aircraft have destroyed our artillery and air defence artillery.'; '241 500 Extremely Urgent From Brigadier General Gelan 17 To Brigadier General Gele Hargheisa 80% of our ration and fuel was destroyed by enemy aircraft. We are very low in supplies. Send what we have requested urgently! 16/8/77 1300.'

102 Former CCS MiG-21 & F-6 pilot, interview granted on condition of anonymity. A confirmation for the loss of both MiG-17s can be found in intercepted SNA Telegram '140 800' provided by Girma: '140 800 Very Urgent From 299 To 11 # TA 45/25/77 The situation is now clear. We have received confirmation the two planes went down around Jijiga area. Both pilots were killed. The locals have buried their bodies. 14/8/77 1000.'

The EtAF rushed both of its C-54s to fly supplies to garrisons isolated by the Somali advance into Ogaden. These slow transport aircraft were highly vulnerable to attacks by CCS MiGs, and required protection by EtAF F-5Es for such operations. (Tom Cooper Collection)

Afework Kidanu flew as wingman to Lagesse Tefera, on 25 July 1977, and then went on to score one MiG-21-kill on 19 August 1977. (via S.N.)

Mengistu Kassaye did not manage to score any kills on 21 August 1977, but claimed at least one under as of yet unknown circumstances, later in the same month. (via S.N.)

Although fresh from a MiG-21-conversion course in the USSR, Ashenafi Gebre Tsadik rapidly refreshed his training on F-5Es and scored one confirmed MiG-21-kill on 21 August 1977. (via S.N.)

Importance of Air Superiority

Shaken by the Somali onslaught, Ethiopian ground troops rallied and, after bringing in reinforcements, on the morning of 18 August they launched a bold counter attack. The EtAF reaction proved crucial, as the ground troops possessed only a few artillery pieces and ground attacks delivering precise hits proved crucial for success in the following battle. At first, two F-5Es established two CAP stations near the embattled town. With Somali air power nowhere to be seen, the fighter-bombers were free to operate against ground targets at will. Flying out of Debre Zeit, almost 400 kilometres away, the EtAF delivered a series of 68 combat sorties, plastering the Somalis with a large number of CBUs and GP-bombs in the process. Canberras joined the fray deploying napalm bombs, and eventually the Ethiopian Air Force was credited with the destruction of sixteen Somali T-54 tanks and up to 80% their other vehicles, enabling ground forces to reorganise and liberate the air base. The fighting for Dire Dawa was especially savage, resulting in extreme losses for both sides. Legend has it that after the battle the few surviving Ethiopian flak-crews found themselves surrounded by piles of empty shell casings and Somali attackers shot them at point blank range. The 15th Somali Brigade later reported suffering over 40% casualties to air attacks alone.[103]

On the morning of 19 August 1977, a single CCS MiG-21MF appeared for the first, and only, time ever above Dire Dawa AB. An

103 Tareke, p. 646 & Babich, *With Foreigners against Our Own*. According to Babich, the Ethiopians deployed not only flaks, but even Soviet-made SA-2 and SA-3 SAMs in defence of Dire Dawa airfield. The latter are known to have a limited surface-to-surface capability, in which they can be fired against larger targets and set to detonate at flight levels of 10–15 metres. Furthermore, Babich emphasised that the counterattack was not launched by Ethiopians but by Cubans driving T-62 MBTs. However, there is no trace of evidence about the presence of anything more than a few dozen Cuban advisers working at Camp Tatek at that time, and even less so about the presence of any of Soviet-made SA-2s or SA-3s in Ethiopia. For a similar reason, Kotlobovskiy's claim that two Somali MiG-21s were shot down by Ethiopian SA-3s while attacking Aisha airfield is without proof. On the contrary, available Ethiopian sources stress that all such reports are fabrications. Their statements are confirmed by the fact that during this first phase of the war the EtAF operated only one early warning radar in Ogaden, the TPS-43 at Karamara Pass, while deployment of Soviet-made SAMs would have enabled them to install additional Soviet-made radars, such as P-12 (ASCC codename 'Spoon Rest'), which were used for that purpose by Somalis. Furthermore, if Dire Dawa was protected by any SAMs, the EtAF would not have had to hold its F-5s temporarily deployed there on permanent standby and ready to scramble at short notice.

Except for AIM-9B Sidewinders (deliveries of more advanced AIM-9Es to Ethiopia were reported but never confirmed), EtAF F-5As and F-5Es also made extensive use of their internally installed 20mm Colt cannons. This photograph shows the installation of the port (right) cannon in the nose of an F-5. (Tom Cooper Collection)

F-5E that took off from Debre Zeit, piloted by Afework Kidanu was vectored by the Karamara radar to intercept, and he not only caught the Somali MiG but also shot it down, scoring his only confirmed kill of the war.[104]

Meanwhile, the EtAF ground personnel scrambled to return Dire Dawa AB to operational condition. Fighting was still raging in the vicinity by the time the runway was cleared of debris on the morning of 21 August. Some fuel, technical personnel, spares and weapons were flown in by C-119s, followed by several F-5s. The EtAF then launched a series of around 50 combat sorties, dispatched at an unprecedented tempo. Most of the F-5 pilots launched six or even seven times that day, Amha Desta no less than eight times. They hit the SNA units so hard, that these not only started falling back, but were forced to withdraw leaving behind a trail of abandoned equipment, burned out tanks, armoured cars, artillery pieces and hundreds of rifles.[105]

While F-5s were flying sorties, several EtAF transports flew supply missions to isolated garrisons elsewhere in Ogaden, and Mengistu Kassaye and Ashenafi Gebre Tsadik, the later fresh back from conversion training on MiG-21s in the USSR, were ordered to provide top cover for them. Due to problems with his canopy hood, Mengistu could not start on time and Ashenafi took off alone. He was airborne for almost 30 minutes before the other F-5E finally arrived. Shortly afterwards the radar station at Karamara advised the pilots of two MiGs heading their way. The situation was now critical, because Ashenafi was left with only 800 litres (1,600lbs) of fuel, while having no choice but to continue the mission and support Mengistu.

As the range to the MiGs decreased, Mengistu advised Ashenafi of visually sighting two MiG-21s on their right side. Remaining at his leader's wing, a second later the Ethiopian wingman sighted another MiG passing low below them, obviously unseen by Mengistu. Ashenafi understood that the third MiG was expected to turn around and attack them from behind. Although feeling uncomfortable, he remained with Mengistu to provide mutual support, while his leader continued following the pair of MiGs that had meanwhile entered a climb. Suddenly, Mengistu called Ashenafi to report that he had lost sight of the Somali fighters; Ashenafi immediately entered a left turn in order to re-acquire the third MiG, which by now had entered a high-speed climb. Approaching his opponent from the side, the Ethiopian fired his gun while passing by, but the closing speed was too high and he missed.

By the time Ashenafi reversed, the Somali fighter was around 2,000 metres (2,187 yards or 6,550ft) away and still heading in same direction as before, and thus in the very centre of an envelope of the Ethiopian's AIM-9 Sidewinder missiles. Obviously, the MiG-21MF pilot never saw his opponent. Ashenafi quickly set up his attack and fired:

104 Ashenafi, interview, 2001; Berhanu, interview, 2006 & Tareke, p. 646. Amazingly enough, some of Russian authors have subsequently claimed that the CCS flew a large number of attacks on Dire Dawa AB and even suffered a number of losses while doing so. Indeed, Babich stressed that Somali air attacks against Ethiopian airfields resulted in Soviet decision to deliver a large number of flaks and SAMs to Ethiopia, including ZSU-23-4 Shilka and ZSU-57-2 radar-controlled SPAAGs. However, Ethiopian participants cannot emphasise strongly enough, that except for this one MiG-21MF, no CCS aircraft ever showed themselves anywhere near Dire Dawa. Similarly, even Somali sources stress that the CCS never flew a single air raid against any main EtAF air base.

105 Ashenafi, interview, 2001 & Berhanu, interview, 2006.

Knocked out Ethiopian Army M41 Walker Bulldog light tank. Some of the last operational Ethiopian tanks of this type were lost during the fighting for Jijiga. (Albert Grandolini Collection)

Company of SNA T-34/85 tanks on parade in Mogadishu, shortly before the Ogaden War. Due to heavy losses of units equipped with T-54s, the older T-34s became ever more important, the longer the war lasted. (Albert Grandolini Collection)

> The MiG did not take any evasive action. Strangely enough, it only flew a few rolls. I believe he did not know about the missile, probably because my radar was not turned on and thus he received no warning from his RWR. The missile at first dropped as if it was heading for the ground, and I was uncertain about what happened. But then I saw the smoke trail coming up and it went straight for the target, cutting the MiG in two. There was a large explosion. The other MiG meanwhile fled back to Hargheisa.[106]

While the surviving MiG indeed disengaged, two other Somali fighters scrambled from Hargheisa. The radar at Karamara Pass thus scrambled two F-5Es from Dire Dawa. The Somalis arrived too late; low on fuel, Ashenafi called his leader to declare his intention to return back to Dire Dawa. Left with barely 200 litres of fuel in his tanks, he attempted to slowly climb to gain some altitude and thus stretch the glide range of his aircraft. Reaching Harar, Ashenafi was at 6,400 metres altitude (21,000ft) when his gauge showed only 100 litres left. Declaring an emergency, he pulled the throttle on idle and then glided straight in to his approach and landing. His engines cut just as the aircraft touched the runway.

High Noon over Jijiga

No sooner had that part of the 26th Somali Division been defeated at Dire Dawa when its other elements opened their assault on Jijiga. Their initial advance was slowed down and held up for nearly a week by the skilful defence operations of the 10th Mechanised Brigade, roughly 40 kilometres east of Jijiga, supported by the 92nd Mechanised Brigade. With the EtAF F-5s and Canberras busy defending Dire Dawa, the CCS MiG-17s and MiG-21s were present in numbers, though most of the time flying only 'hit-and-run' attacks.

By the end of August, Jijiga turned into the site of the most bitter struggle of the entire war. Determined to take the town at any price, the Somalis deployed all of their firepower to overcome the defences. Their efforts culminated on 1 September, when the Ethiopian defenders concluded that further resistance was futile and withdrew in the direction of Adew and the Karamara Pass.

Aerial fighting during the battle for Jijiga was also intensive. Early in the morning of 1 September, Lagesse Teferra and Ashenafi Gebre Tsadik were underway from Dire Dawa with the task of attacking one of the Somali columns, when the radar at Karamara advised them about the appearance of two MiG-21s that were heading in a similar direction. The Ethiopians immediately made a combat turn and entered the 'combat spread' formation, establishing a separation of around 1,500 metres (roughly 5,000ft) between each other. While approaching their target, they made several cross-turns in order to maintain situational awareness and check that no enemy fighters were behind them. Ashenafi recalled:

> Lagesse spotted the MiGs first, but shortly afterwards I lost the sight of him. Expecting there would be a third MiG around, I entered a manoeuvre designed exactly with this kind of situation in mind. Called the 'egg', this manoeuvre consisted of a semi-loop at acceleration of 4g, so that I would be able to observe the situation below and above me, while also capable of breaking the lock-on of Soviet-made R-3S air-to-air missiles, which can only track targets turning at 3gs or less. Mid-way through this turn, I sighted the third MiG, circling around 1,000 metres below and waiting for me to make a mistake. I dived directly in his direction. The Somali saw me and panicked; he evaded by entering a climb but in this way exposed his hot engine exhausts to my Sidewinders. The F-5E is far superior to the MiG-21 in turning combat. Making use of the better turning rate of my aircraft, I curved behind the target and fired a missile. The Sidewinder went astray as I was too close. That was just as well as I was so close, if the missile had hit, I would have been caught in the debris of the MiG and got killed in the process. Moments later I found myself in a scissors manoeuvre with the Somali fighter. We crossed above each other three times, and each time I fired a short burst from my 20mm cannons. The Somali did not roll out from the third roll, but went down and hit the ground. It is possible that one of my rounds hit his cockpit.'

In the meantime, Lagesse engaged another MiG and the two fighters entered a wide turn. After gradually gaining advantage over his opponent, and with the Somali clearly away from the sun, the Ethiopian then fired an AIM-9 Sidewinder. The missile cut the turn and impacted, causing a big orange and black fireball. Lagesse Teferra could not have known this at the time, but this was the EtAF's 11th – and final – kill officially confirmed as an air-to-air victory of this war.[107]

By 1 September 1977, the CCS is known to have suffered a

106 Ashenafi, interview, 2001.

107 Ibid.

Taken at the same time, this photograph shows the SNA's BTR-60 armoured personnel carriers that were deployed by four mechanised brigades during the Ogaden War. (Albert Grandolini Collection)

confirmed loss of 9 MiG-21MFs and 2 MiG-17s to Ethiopian fighters as well as at least three additional losses caused by crashes of its pilots during air combats. The total number of Somali fighters lost so far increased to twelve MiG-21s and thirteen MiG-17s, or almost half the entire pre-war force. Indeed, several CCS pilots are known to have suffered the unpleasant experience of being shot down by Ethiopians several times (one of them later recalled about a colleague of his, and one of best Somali MiG-17-pilots, Capt Hussein Hassan Abuker, who was shot down no less than three times during the first two months of the war).[108] Unsurprisingly considering this heavy attrition rate, the CCS never challenged Ethiopian aerial superiority over Ogaden again. Although this did not mean that the EtAF would not keep on scoring, or that the tide in the war generally turned against Somalia, in regards of the battle for air superiority there was an undisputable winner. Over time, this was to prove decisive for the overall flow of the conflict too.

Loss of the Ace

With 4½ officially confirmed kills to his credit, and two not accepted as such by the EtAF, Lagesse Teferra was now the leading Ethiopian fighter pilot. However, he was about to run out of luck on 1 September 1977.

Returning to Dire Dawa around noon, Lagesse was ordered to provide top cover for a Canberra flown by Mesfin Haile on a bombing mission in the Balli region, a few kilometres inside Somalia. Lagesse flew ahead of the bomber at low level to scout the route towards the target. Underway over an area said to be under the control of the Ethiopian Army, he stumbled over advancing Somali units and his F-5E was hit by flak (probably by a ZSU-23-4 Shilka SPAAG). Lagesse was lucky, for his airplane remained controllable for a while and he managed to get some distance from the site of this encounter before ejecting. Propelled out of the cockpit below the minimal recommended level he was also lucky to survive this action. Nevertheless, neither his wingman, nor the Canberra crew, or the Somalis that shot him down, saw him ejecting.

Lagesse landed in the Negele Borena area, close to the Wabe River in the middle of a thick forest. His situation was extremely dangerous: the river was full of crocodiles and pythons, lions roamed the area, and the local population was not in the least friendly. After finding a place not infested with wild animals, he climbed up a tree (lions seldom do the same due to the thorns). About an hour later, he saw an EtAF SA.316B helicopter passing by and fired a flare, but this obviously went unnoticed. The next morning, Lagesse was awaken by human voices, which turned out to be women going to fetch water. He jumped from the tree, scaring them, but they offered him water and took him to their village, eventually providing him with some food as well. What the Ethiopian did not recognise right away was the fact that these were the wives of WSLF guerrilla fighters, allied with the Somalis. While feeding him, they also sent a messenger to notify his enemies. Lagesse was arrested a few hours later, without any incident. He handed over his pistol and then waited until Somali Col Abdullahi Yussuf Ahmed, the CO of the region, showed up in a Land Rover. Abdullahi treated the downed Ethiopian officer very well, and even let him sleep in his tent. Later

108 Former SAC MiG-21 & F-6 pilot, interview granted on condition of anonymity. According to same source, Capt Abuker was then shot down for the fourth and sadly final time, see Chapter 5.

The ubiquitous SA.316B Alouette III helicopters of the 14th Squadron EtAF were few in number but often played an important role. (Hervé Dessallier via Albert Grandolini)

on, Lagesse was taken to Baidoa and then flown to Mogadishu. He spent the next eleven years at the Lanta Borre prison in Mogadishu, together with a number of Somali political prisoners.

The EtAF learned about Lagesse's capture only through intercepted Somali radio messages. The person that first heard a corresponding message was a senior EtAF mechanic of Somali descent, who later became a mayor of Debre Zeit. Lagesse was exchanged for Somali PoWs following the reconciliation between Ethiopia and Somalia in 1988. He was promoted to the rank of major general on return to Ethiopia.[109]

Fall of Karamara Pass

The success of the EtAF, but also the loss of Lagesse Teferra, went completely unnoticed by the public, and were barely felt by the Ethiopian military. Namely, the defeat at Jijiga caused such a shock that, like in a well-orchestrated drama, Mengistu Haile Mariam flew to Harar and, in a daring personal act, organised a counterattack, leading his men into the battle himself. Jijiga was temporarily recaptured by a motley collection of Ethiopian Army and militia units in the course of two assaults and very bitter fighting on 5 September, but the SNA counterattacked, eventually putting the Ethiopian units inside the ravaged town under siege. The leading Derg then escaped to Adew, leaving his troops to falter again in the face of renewed Somali attacks, losing not only some 75 tanks (including the last operational M41s) and 71 APCs, but also plenty of arms, ammunition and supplies in the process.[110]

Over the next few days there was a lull in fighting as the CO 26th Division SNA, Maj Gen Mohammed Nur Gallel, was forced to reorganise his battered units before launching a pursuit of the retreating Ethiopians. Following the policy of total commitment of all available resources introduced right from the start of this war, Gallel demanded the CCS to support this operation, but the Somali Air Force was only sporadically able to do so. While the units of the 26th Division restarted their advance around the 5 September, it was not until four days later that two CCS MiG-17s were sighted by the Ethiopian military. Their attack on one of the retreating Ethiopian columns was rudely interrupted when one was hit by ground fire and shot down on Mount Anbessa. The pilot ejected safely and fortunately for him, an attempt by EA units to find and catch him ended in a short fire fight with SNA troops, in which three Ethiopian soldiers were killed.

With Ethiopian units falling ever further back, the SNA advance then picked up speed and on 12 September the Somalis achieved two major feats. During the day, their long-range artillery knocked out the TPS-43 radar at Karamara Pass, severely hindering the effectiveness of future EtAF operations. This loss forced the air force to dismount its second system of this type from Debre Zeit and re-locate it to Mount Megezez, on the road from Addis Ababa to Debribirhan, from where it was unable to provide as good coverage

109 Col Abdullahi Yusuf Ahmed continued a successful career with the SNA until distancing from Barre and emigrating to Ethiopia, in 1979. He returned to Mogadishu in 2007 and served as President of the Transitional Federal Government of Somalia, from 2004 until 2008.

110 According to Tareke (p. 648), documents from the Ethiopian MoD show that Mengistu was in Harar at that time, although some observers express doubts if he personally led the counterattack on Jijiga. According to Babich, the fall of Jijiga occurred after a nocturnal attack by a 'squadron of Somali Il-28 bombers'. Supposedly, the unexpected raid by night caused such a panic, that the Somalis experienced only minimal problems while finally securing the town.

of the skies over Ogaden.

Furthermore, in the evening, Somali units then successfully assaulted the Karamara Pass, capturing it from the units of the 2nd Militia Division. Reinforcements in the form of an entire Ethiopian armoured brigade deployed form Addis Ababa towards Harar, as well as several artillery detachments, arrived too late to prevent the local Ethiopian commander from disobeying his orders to hold this position at any price.

The defeat at Jijiga and loss of the Karamara Pass marked the lowest ebb of the war for Ethiopia. The Somalis not only delivered a crushing blow to the Ethiopian military, but also in terms of psychology and capabilities, the Ethiopian Army was in possession of no resources and no equipment that would enable it to re-conquer the area. In desperation, the Derg felt compelled to call for a general mobilisation, pleading to the patriotic feelings of all Ethiopians. Meanwhile, surviving elements of the 3rd Division were ordered to gather at a point some 50km east of Harar and hold out to the last man, if necessary. The Ethiopians could not afford to withdraw any more.

Battle of Harar

What saved Ethiopia at this critical moment in time was the Somali failure to exploit an advantageous situation, primarily caused by the continuous and heavy EtAF air raids on its supply convoys. Initial SNA operations in northern Ogaden were undertaken relatively close to bases and depots in northern and western Somalia. However, their advance eventually brought the Somalis into areas predominantly populated by hostile Ethiopian ethnic groups and ever closer within the range of the Ethiopian Air Force. Their supply situation began to worsen with every day, and then with every hour. Exhausted from heavy fighting and with low stocks of ammunition, fuel and food, the SNA could not launch a pursuit of the Ethiopians in the direction of Harar, and instead satisfied itself with the fortification of newly-established positions. Thus, it was not before October that the SNA found the strength for another offensive.

This break in the fighting enabled the Derg to find a solution for the major problem they were now facing. Although the EtAF had established air superiority over Ogaden, the army was badly mauled and, even when combined with militia, hopelessly out of condition to re-conquer the area. Plenty of new arms, preferably acquired at no or very low cost because the nation was bankrupt, would be required even to reinforce the military to a point where it could effectively resist any further Somali attack. Some help had by that time arrived from South Yemen, which in late September deployed two armoured battalions equipped with T-34s to Ethiopia. But this force was barely sufficient to help the defence of Harar, which was meanwhile encircled by Somali forces from three sides, and practically under a siege.

The defence of this town with a population of around 48,000, and the base of the prestigious Ethiopian military academy, was organised by the remnants of the 3rd Infantry Division. This included a conglomerate of mechanised battalions and companies, reinforced by the 75th Mechanised Brigade, the Kagnew Brigade (remnants of the 76th and 96th Brigades), 2nd Tank Battalion, 219th Nebelbal Battalion, 4th Air Defence Battery and several battalions of poorly armed reservists. On the other side, the SNA mustered five motorised infantry brigades for its attack, supported by a full tank brigade and a reinforced artillery brigade, in addition to several SALF battalions and one commando brigade.

The Somalis first attempted several frontal assaults, through late

While slightly inferior in manoeuvring performance to the F-5E, F-5As acted as the primary Ethiopian fighter-bomber. Armed with general-purpose bombs and 2.71in unguided rockets, as well as internal 20mm cannons, they contributed significantly to the overall success of the EtAF during the Ogaden War. This photograph from the early 1970s (note the old IEAF roundel) shows the third F-5A delivered to Ethiopia. (Albert Grandolini Collection)

September and October, aiming for a concentration of Ethiopian forces in a bulge that expanded from Harar south-east towards the town of Kore. An attempt to outflank the 92nd Mechanised Brigade at Gursum failed though, and the fighting degenerated into bitter trench warfare for Mount Dalcha, a few kilometres south of Kore. Mount Dalcha changed hands several times and combined casualties on both sides probably exceeded 3,000; the final Somali assault on this hill took place on 19 October and was repulsed by the newly-deployed 74th Ethiopian Brigade. The Somalis lost 219 KIA and two MiG-17s shot down in the process, according to Ethiopian sources.[111] Two days later, another MiG-17 was shot down by Ethiopian flak over Harar, and this time Capt Hussein Hassan Abuker ran out of luck; he was killed after failing to eject from his burning aircraft.

The last Somali offensive of the war began on 23 October, when the SNA opened an attempt to capture Jarso, 35km north-west of Harar. The Ethiopians rushed the Kagnew Brigade and a battery from the 3rd Artillery Brigade to the scene, causing over 600 casualties on the other side and stopping the attack for the time being. On 30 October, another Somali assault hit the Ethiopian frontlines at Fedis, where two paracommando battalions and the 501st Militia Brigade, supported by an artillery battalion and a platoon of M41 tanks, resisted for nearly a week before giving up. By 4 November, Harar was thus in serious danger of being overrun by advancing Somalis and only the arrival of the 2nd Paracommando Brigade from Addis Ababa prevented a potential disaster.[112]

After licking their wounds and collecting whatever supplies they could, the Somalis re-attacked Jarso on 16 November, this time with intensive artillery, and even some air support. Extensive deployment of BM-21 MRLs collapsed the Ethiopian resistance and the SNA finally broke through in the direction of Kombolcha, only 16km north-west from Harar. Although most of the panic-stricken defenders abandoned the latter town, timely intervention of the 1st Ethiopian Paracommando Brigade saved the situation and this prong of the Somali advance was stopped too.

Except for deployment of elite paratrooper units, another impulse for successful Ethiopian resistance was again provided by the EtAF. On 16 November, a combined formation including two F-5Es, two F-5As and two Canberras hit the Somali forward base at Jeldessa, east of Dire Dawa. What the Ethiopians did not know was

111 Tareke, p. 653.
112 Ibid., p. 654.

A jubilant Mengistu together with Raul (left) and Fidel Castro (centre) during a reception for Cuban dignitaries at Bole IAP. Mengistu's skillful manoeuvring on the diplomatic scene attracted not only Cuban, but eventually also Soviet support for Ethiopia. (Albert Grandolini Collection)

Siad Barre's manoeuvring on the international diplomatic scene was not as skilful as that of Mengistu, or at least Somalia did not appear as an attractive ally to most foreign observers. Although visiting not only Moscow, but also Romania (this photograph shows Barre with Romanian strongman Nikolai Ceausescu), he failed to attract the necessary support. (Albert Grandolini Collection)

that the SNA had its main ammunition depot there and that because of this, the area was heavily protected by flak, including several ZSU-23-4s, which EtAF pilots considered their most dangerous opponents. Piloted by Berhanu Kebede and Tesfu Desta, the two F-5As were the last to make their attack, diving from an orbiting position of almost 6,000m (18,000ft) high. While approaching this target, Berhanu's aircraft received a heavy blow to the left side of front fuselage, which ripped away the radome, port 20mm cannon and radio, the debris of which ruined the left engine as well. The pilot regained control of his aircraft and considering his situation, made a turn for Dire Dawa. Everything functioned fine until one of the landing gear struts failed to lower. If he could have reached Debre Zeit even this would have been no particular problem, but Dire Dawa lacked the arrestor barrier installed at the end of runway with the purpose of stopping damaged aircraft. Eventually, Berhanu was forced to eject near the air base, landing under his parachute on the top of a cactus tree. He was picked up by an EtAF Alouette helicopter only minutes later.

Despite the loss of a precious F-5A, the Somali position was not only hit by Ethiopian bombs, but literally blown away during this attack. Unsurprisingly considering this important loss, the Somali offensive was finally stalled and the SNA was never to recover. Mogadishu thus missed its final opportunity to take Harar and just a few days short of the war entering its decisive phase.

CHAPTER 5
SWITCHING ALLIANCES

Desperate to obtain weapons that would enable its army to recover Ogaden, the Ethiopian government was meanwhile demanding aid from the USA. With Washington turning all requests down, the Derg decided to establish a closer relationship with Moscow. Following a visit by Cuban leader Fidel Castro to Addis Ababa, Mengistu went to Moscow in March 1977, only to realise that the Soviets were not keen to cooperate with him without a clear sign of Ethiopian friendship. It was in an attempt to 'deliver the message to Moscow', that Mengistu ordered the closure of all the remaining US military installations in Ethiopia (Asmara and Masawa), and requested the MAAG team to finally leave Addis Ababa in April 1977. When this was still not enough, Mengistu went a step further and during the next visit by Castro on 16 April 1977, declared himself ready to unite Ethiopia with Djibouti and Somalia in a 'Federation of East Africa'; for several years this was a long-term Soviet aim for this area. Upon return from another visit to the USSR in May 1977, the Ethiopian leader ordered the closure of the offices of Military Attachés to the Embassies of the USA, Great Britain, and Egypt.[113]

113 According to Feleke (in *It Happened Like That*), there is a funny anecdote about how Mengistu Haile Mariam finally convinced Moscow to start selling armament to Ethiopia. During one of his trips to Moscow, he was coldly received and had to realise that the Soviet leadership was not keen to even see him. After meeting KGB officials and different generals for days without any results, finally, Mengistu was offered to lay out his case to Soviet leader Leonid Brezhnev during a dinner. The latter is said to have replied, 'Comrade Mengistu, I think you need rest. Why don't you relax at the Black Sea for a few days and then we will discuss your request?' Frustrated, Mengistu got up from the table and stormed out, snapping at Brezhnev that he did not come for vacation while his nation is in peril. The Soviets were shocked by his behaviour but this incident made the seriousness of the issue clear to them. Shortly later, Brezhnev turned to Mengistu Gemechu, Haile Mariam's Chief of Staff and senior Derg member, advising him (through a translator), 'Comrade Mengistu is a revolutionary, but not a socialist. He will first have to correct that.' Mengistu, who was no leftist by political orientation, became one because he did not have any other choice. He subsequently declared himself as a communist, sealing his political future, but also that of Ethiopia for the next thirteen years and beyond. In return, he was granted permission to order MiGs, just for the start.

Bole IAP in late November 1977, with two V-TA An-12s dwarfing an Ethiopian Cessa 185 (foreground) and an EtAF Dove (centre, behind the truck). For several weeks, a Soviet transport would land at this airport every 20 minutes. (Albert Grandolini Collection)

Moscow reacted almost immediately this time, and 13 contracts for armament worth US$400 million were signed between Ethiopia and the USSR only a few days later. As described in Chapter 3, these were primarily related to no less than 48 MiG-21bis, but also envisaged deliveries of six brand-new Mil Mi-24 helicopter gunships, 12 Mi-8 transport helicopters and more than 300 MBTs with assorted munition for the army. The plan for delivery of MiG-21s was thus set several months before the Ogaden War, and this was the reason for so many EtAF pilots travelling to Kusovskoye in the USSR for conversion courses, early during the war. While few of them returned during the following months, the others completed their training and thus the EtAF was ready to press its new aircraft into service as soon as these were delivered, in November 1977.

Meanwhile, Mengistu reached another agreement with Fidel Castro, which envisaged the deployment of a Cuban contingent including MiG-17s and MiG-21s, starting in October 1977. This was to become involved in the process of re-training Ethiopians on Soviet equipment, and then to help them drive the Somalis out of Ogaden.

Barre's Crucial Mistake

The President of Somalia was also involved in intensive negotiations with Moscow, requesting deliveries of additional weapons and stocks of ammunition and supplies. As a declared Soviet ally and personal friend of Leonid Brezhnev, who decorated him with the Soviet Medal of 'People's Friendship', he was repeatedly disappointed by the Soviets turning all his requests down. On the contrary, Brezhnev was annoyed by Barre's refusal to negotiate a peaceful solution to the issue of Ogaden and unite Somalia in the 'Federation of East Africa', but also his false promise that Somalia would never attack Ethiopia.[114] Desperate, the Somali leader eventually entered negotiations with the USA, and on 26 July 1977, the State Department announced that the USA, Great Britain, and France were prepared to provide arms to Somalia. However, this decision was, at least unofficially, reversed in light of the invasion of Ogaden, and no armament or equipment were furnished for the time being.

Concerned about announcements of Soviet deliveries of MiG-21s to Ethiopia, Barre next attempted to make sure that the USSR would at least remain neutral regarding the Ogaden War. Correspondingly, he visited Moscow in September 1977, but only to receive a negative response. On the way back to Mogadishu, the Somali President made a stop in Cairo to meet the Egyptian President Anwar as-Sadat and sign an agreement for military assistance worth approximately US$30 million.[115] Meanwhile, all the Somali attempts to obtain arms from Iran and Saudi Arabia failed to produce results, even though Iraq eventually signalled preparedness to deploy twelve instructor pilots, together with some fuel and non-lethal supplies.

Ultimately, the situation became completely untenable for Somalia. Following another visit to Moscow in early November 1977, during which he was informed that Brezhnev was on vacation in the Crimean Peninsula and thus out of reach, Siad Barre lost his temper. Perhaps following the example of Egyptian President Sadat in 1972, he made a crucial decision: on 13 November 1977, Barre terminated the 20-years Treaty of Friendship with Moscow and demanded all the 20,000 Soviet specialists and military advisers to leave within one week.[116]

Hostile crowds menaced the 1,500 departing advisers and their families as they clambered aboard the Soviet amphibious ship *Krimsky Komsomolets*, in the port of Berbera, hastily taking with them all their belongings. In every sense the departure was undignified, particularly because the crew of the ship was about to open fire on Somalis in order to force entry into the harbour. The situation in Mogadishu was even worse. There the Soviets not only had to deploy an entire squadron of their warships to enable amphibious ship *50 Let Sheftsva VLKSM* to enter, but also deploy two companies of marines on the quay, to prevent their citizens from being lynched

114 Babich, *With Foreigners against Our Own*.

115 Aidid et al., Chapter 14 & Babich, *With Foreigners against Our Own*. Interestingly, while it is usually said that this order was secretly paid for by the Saudis, who were always keen to 'return' Somalia to the West, Aidid, who detailed that this contract included only spares and ammunition for T-54 MBTs and various infantry-related equipment, claimed that Somalia paid for Egyptian arms through credit. Furthermore, except for Babich, there is no other source that could confirm any supposed Egyptian agreement to sell MiG-21s and MiG-17s to Somalia. On the contrary, most Egyptian MiG-21s operational around this time are accounted for. Although nothing similar can be said for their MiG-17s, of which there is no trace of their delivery to Somalia either.

116 Mezentzev, *Ethiopian-Somali Border War*, p. 32.

Soviet arms enabled Cuban advisers to train additional thousands of Ethiopian militiamen, necessary for launching a major effort to re-conquer Ogaden. This photograph shows a company of militiamen on parade in Addis Ababa, before they went to Ogaden. (Albert Grandolini Collection)

In addition to enough firearms and ammunition to establish two 'Militia Armies', the Soviets also significantly bolstered Ethiopian air defence capabilities by delivering SA-2 and SA-3 SAMs, the latter of which are seen in this photograph of a post-Ogaden War parade. (Albert Grandolini Collection)

This ex-Soviet Air Force and ex-People's Democratic Republic of Yemen Air Force (PDRYAF) MiG-15UTI was transferred to Ethiopia, together with 12 MiG-17Fs from PDRYAF stock, in November 1977. Considering them inferior to their requirements, the Ethiopians refused to accept these aircraft and thus, although they received EtAF serials and national markings, they were flown exclusively by Cuban pilots during the Ogaden War, and stored immediately afterwards. (via Pit Weinert)

Conversion training of Ethiopian pilots on to Soviet-made aircraft generally proceeded at quite a high pace and without any major problems. Here a flight of EtAF MiG-21bis's as seen shortly after delivery. (EtAF)

by the mob.[117]

Political Manoeuvring

It is hard to say to what degree Siad Barre was aware of the consequences of his decision for Somalia and for other countries. In strategic terms, the USSR lost the use of the extensive port of Berbera and four large airfields it had constructed in Somalia. But, in turn it replaced the US presence in Ethiopia, which previously had cancelled all of its ties to Washington. Indeed, while the two amphibious ships brought all the Soviet citizens evacuated from Berbera and Mogadishu to Aden, in Yemen, many of the evacuees immediately received an order to continue their journey straight to Ethiopia.

The Somali action definitely offended Moscow beyond the point where Brezhnev would care about the possibility of a confrontation with the USA, Saudi Arabia, Egypt or anyone else. Brezhnev decided to make an example of Somalia and send a strong signal to all of the other Soviet allies that might prove as unreliable.

The Soviets began their preparations for deploying advisers to Ethiopia on 13 August 1977, when their government issued the Decree 1823, according to which the military was to form a group of advisers led by Lt Gen P T Chalplygin. However, these did not arrive in Ethiopia before the Cuban contingent of around 2,000 officers and other ranks under General de Division Arnaldo Ochoa Sanchez in early November.[118]

Subsequently, the situation experienced a fundamental change. Within only a few days of the Soviets being forced to leave Somalia, the number of their advisers in Ethiopia jumped to 1,500. The entire contingent was put under the command of Lt Gen Vasili Ivanovich Petrov, then the Deputy C-in-C Soviet Army. Around 100 Soviet Air Force (V-VS) officers under the command of Lt Gen G Dolnikov was delegated to the EtAF too, though most of them arrived only in December. Most importantly, starting on 25 November 1977, the Soviet Military Transport Aviation (V-TA) initiated an intensive air-bridge to Ethiopia, delivering huge amounts of armament and material. For this purpose, the V-TA concentrated no less than 225 Ilyushin Il-18, Antonov An-12, and Ilyushin Il-76 transports on airfields in Tashkent, where these were loaded with a total of around 600 T-55 and T-62 MBTs; 300 BMP-1, BTR-60 and BRDM-2 IFVs and APCs; 400 artillery pieces; several dozens of BM-21 MRLs, and enough ammunition and supplies for three small divisions. The air bridge went from Tashkent via Baghdad (for a refuelling stop), over the Persian Gulf and Saudi Arabia to Aden and Masawa, and then to Addis Ababa. For six months, in a unique development that stunned most foreign observers, a Soviet transport aircraft was landing at

117 Babich, *With Foreigners against Our Own* & Mezentzev, *Ethiopian–Somali Border War*, p. 31. Both Soviet amphibious ships used for this evacuation belonged to the Project 1171/Tapir class of landing ships (ASCC-designation Alligator).

118 Urban, p. 44.

Notable on this An-24 is the 'Flying Leopard' emblem applied on both sides of the front fuselage of many Somali Air Force aircraft. Except in their primary role, as transports, CCS' An-24s were several times deployed as bombers for nocturnal operations, using bomb-racks and NKBP-7 bombing sights. (Albert Grandolini Collection)

Bole IAP outside Addis Ababa every 20 minutes, by day and by night. About 50 Soviet warships and merchant ships were involved in transporting T-54 and T-55 MBTs, artillery (including 130mm M54 cannons and 122mm D-30 howitzers), air defence systems and assorted ammunition from the port of Sevastopol, through the Suez Canal towards the port of Asseb by April 1978.[119]

Ethiopian MiGs and Foreign Pilots

With the help of Soviet equipment delivered during November 1977, the Ethiopian Army kitted out 100,000 newly trained troops. About 30,000 of them, organised into the 1st Revolutionary Liberation Army, were almost immediately deployed to Ogaden.[120] The re-equipment of the EtAF proceeded at only a slightly slower pace. During November, the Soviets delivered a total of twelve or thirteen MiG-17Fs and two MiG-15UTIs (all from the stocks of the People's Democratic Republic of Yemen Air Force), eight newly-built MiG-21UMs, twelve MiG-21MFs, and twenty Mi-8Ts. Except for the MiG-17s and MiG-21MFs, which were donated to the EtAF by Moscow, all the other aircraft were officially ordered by Addis Ababa and as such paid for by Ethiopia.

The first Soviet aircraft activated in Ethiopia were MiG-21UMs, which entered service with the re-activated 33rd Squadron in order to convert ex-F-86-pilots of the 1st and 2nd Squadrons EtAF. This process was supported by a group of Soviet advisers, primarily pilots drawn from 160 Fighter Regiment (then based at Borisoglebsk AB) and 927 Fighter Regiment (Byeroza AB) V-VS, that arrived at Debre Zeit AB in December 1977. Most of them remained in Ethiopia for barely a month and all had left by July 1978. Even though several EtAF pilots, including Berhanu Wubneh, were qualified on MiG-21s in time to fly their first combat sorties in early January 1978, overall, the process of re-equipping two squadrons with MiG-21s was not completed before the end of the Ogaden War. Indeed, most of the re-equipped EtAF squadrons were officially declared operational on Soviet aircraft only later in 1978.

The detachment of the Defesa Anti-Aérea y Fuerza Aérea Revolucionaria (DAAFAR, Cuban Air Defence Force and Air Force) also arrived at Debre Zeit AB in November 1977, but this was completely different than usually reported. Led by Lt Col Ruben Iterian, it consisted of only a small group – between 8 and 10 – pilots at most, that had previously served as advisers in Yemen. Iterian and his men were originally supposed to help convert EtAF pilots to MiG-15UTIs, MiG-17Fs and MiG-21MFs. However, after testing these aircraft, the Ethiopians concluded they were useless for their purposes, and continued insisting on deploying slightly more advanced MiG-21bis's instead. Although the Soviets protested against such demands, reasoning that the MiG-21bis was by then exported only to one foreign ally, East Germany (which was not entirely true), they eventually agreed and provided 48 aircraft of this variant, starting in December 1977.

Iterian agreed for his pilots to fly combat sorties on the EtAF side, but insisted they serve in a self-contained unit, not flying joint operations with the Ethiopians because of language barriers. A decision was eventually taken to organise the Cubans as a provisional 4th Squadron, equipped with a mix of MiG-15UTIs, MiG-17Fs and MiG-21MFs. A few additional Cubans might have flown some of the aircraft assigned to the 14th Squadron, but Russian reports about Cubans flying all the MiG-21Rs and Mi-24s delivered in December 1977, cannot be confirmed. The Ethiopians do confirm to have ordered four MiG-21Rs and up to 28 Mi-24s (as well as 30 additional Mi-17s instead of older Mi-8s), and then the first batch of 44 MiG-23BNs (out of an eventual 100) by the late 1977, but according to them, except for MiG-23BNs, none of these aircraft entered service before the end of the Ogaden War. Indeed, the first group of EtAF Mi-24 pilots, including Berhane Meskel, was still undergoing their conversion training in the USSR when the war ended. Correspondingly, the Cubans were operating on their own

119 Ibid & Mezentzev, *Ethiopian–Somali Border War*, p. 36; citing other articles by Soviet veterans of the Ogaden War, Mezentzev specified that V-TA's An-12s 'operated a total of 2,671 flights and transported about 17,000 military personnel, more than 1,000 wounded personnel, and 14,363 tons of cargo, including 4,059 tons of military equipment and ammunition. Furthermore, the Soviets went a step further and stretched this air bridge to the other side of Africa as well. Deploying all 18 Antonov An-22 Antey transports then in service with the 8 and 81 Regiments of the V-TA, they began flying heavy equipment via Addis Ababa to the Cuban contingent in Luanda, in Angola, as well. Eventually, the An-22s delivered a total of 455 tons of various supplies as well as 37 heavy vehicles to Angola, but their operations are not to be misunderstood as related to developments in Ethiopia.'

120 Tareke, p. 656.

With their T-54s becoming one of favourite targets for Ethiopian air strikes, this crew of a Somali Army T-54 took care to conceal their vehicle under a tree during a break in fighting. (Albert Grandolini Collection)

for most of the war while no Soviets ever flew any combat sorties.[121]

Raid on Berbera

The arrival of Soviet advisers that used to serve in Somalia did provide the Ethiopians with some particularly useful intelligence, especially so thanks to Col Gen Borisov, who until 13 November commanded the Soviet mission in Somalia and was subsequently assigned as deputy to Petrov. Borisov brought with him several other officers with extensive experience from Somalia and an intricate knowledge about the SNDF, including Col Netchanov, Lt Col Andrey Filatov, and Lt Col Semyon Nezhinsky. Furthermore, on 4 December 1977, the Soviets launched their reconnaissance satellite Kosmos 964 into an orbit over the Horn of Africa, in order to obtain better intelligence about the SNA's battlefield dispositions and movements, and this satellite is known to have arrived on 17 December, delivering timely information.

In the case of the EtAF, the intelligence collected in this fashion enabled Ethiopians to learn important details about Hargheisa and Berbera ABs. which in turn gave the pilots of 9th Squadron an idea. With a detailed layout of the two main Somali air bases in their hands, a group of pilots led by Techane Mesfin were in a position to not only draw up a very precise plan for attack on these installations, but also train accordingly.[122]

Flown on 27 December 1977, the raid on Berbera AB was undertaken by four F-5Es, led by Techane, and supported by one C-119 that acted as an airborne command post. Each Tiger II was armed with a pair of Mk. 83 500kg bombs, and a full load of 20mm ammunition, in addition to wing-tip mounted Sidewinders. The Ethiopians first cratered the runway and then two of them, led by Bacha Hunde, turned around to strafe. Bacha Hunde managed to destroy the last operational Somali Il-28 bomber on the ground, while his wingman damaged the MiG-21MF, serial number '261'.[123] Colonel Netchanov later recalled to what degree this operation emboldened everybody involved:

> After the successful attack on Berbera, we decided to hit some other places, including the local port. It was maddening that all these things were reconstructed and expanded with Soviet money only a few years earlier, and now we expected them to fall into American hands. We had no doubts that the Americans would soon be there, because we did not only depend on news from newspapers. Considering the fact that the Somali air defences could not put up any meaningful resistance, I prepared a provisional plan for this attack and submitted it to Lt Gen Dolnikov. He listened carefully, and asked few questions, then granted his permission for development of a full plan.
>
> During our subsequent private conversation, he let me know that a contingent of two or three Cuban squadrons would soon arrive, and thus considerably reinforce the Ethiopians. I was to take that development into consideration. After the briefing, Dolnikov also issued an order to consider all the eventualities and let only Soviet instructors work on that plan. He intended to reveal its details to the Cubans and Ethiopians only the evening before it was to be launched. We worked on that plan by night and in our free time. All documents and calculations remained strictly within our headquarters, under the guard of Soviet airborne troops that were protecting us.
>
> As ever more Cuban units arrived, they became part of the plan and we increased the number of targeted objectives. Considering all the details known to us, we did not expect anything to be left standing after that attack. As Gogol's Taras Bulba would say, 'I created you and I am going to destroy you'.

Supposedly, when Dolnikov submitted this plan to Moscow for final approval, somebody up the chain of command decided that this was demanding too much. Netchanov and his colleagues were ordered to send all the materials to the USSR and then forget about them. Unaware of political manoeuvring within the framework of the Cold War, they experienced this negative response as a 'cold shower'.[124]

Actually, it seems the EtAF did not consider the Soviet plan technically feasible. By this time, and following the loss of Lagesse and Afework, the Ethiopians were left with only four F-5Es and

121 Berhanu, interview, 2006; Tom Long, interview, Oct. 2003; Babich, *With Foreigners against Our Own*. Note that reports about deployment of East German military advisers, including pilots for Ethiopian fighter-jets, have proved wrong. As detailed by Möller in *DDR und Äthiopien*, pp. 4–6, former East Germany limited its engagement to deliveries of AKM machine guns, ammunition and several thousand trucks. Ironically, various Russian sources meanwhile make a lot out of emphasising that both the EtAF and the DAAFAR were practically put under the command of Gen Dolnikov, and that Dolnikov commanded all of their operations for the rest of the Ogaden War and that Soviet pilots flew combat sorties as well. Not only Ethiopian, but also Cuban sources deny any such claims and there is no evidence of any kind of participation of Soviet pilots in the Ogaden War. Furthermore, although under a unitary command, due to language differences, the EtAF and DAAFAR were only seldom able to fly joint operations. Indeed, because most of the involved Soviet officers spoke no English, which was practically the official language of the EtAF, and no Spanish either, and their operational procedures were entirely different than those of the Ethiopians, the Ethiopians have concluded that it would be entirely irresponsible to let any of them fly operational sorties over Ogaden.

122 Babich, *With Foreigners against Our Own* & 'List of Kosmos Satellites', *Wikipedia*.

123 Interviews with Ashenafi and Berhanu, and with former CCS MiG-21 & F-6 pilot (the latter granted on condition of anonymity); additional information about the effectiveness of this raid was provided by Girma, in the form of Somali telegrams '271 725' and '28/12/77 1100', intercepted by Ethiopian military intelligence: '271 725 Very Urgent From Hargheisa Air Base To Mogadishu Air Base Q/CC/X/KT/1/9-1188 Today at 1410 we were attacked by enemy aircraft. One Il-28 was destroyed. A MiG-21 with tail number 261 was also damaged. Lt Usman Sheik was wounded in his leg. The planes only attacked the air base. Major Garan 27/12/77 1745'; 'To Mogadishu People are fleeing to Mogadishu Yesterday's attack by enemy aircraft has resulted in damages. They dropped bombs in five places. Our weapons depot at Iedi was completely destroyed. 28/12/77 1100.'

124 Babich, *With Foreigners against Our Own*.

Although of poor quality, this rare photograph shows one of two Canberras, either 351 or 354, that remained operational with the EtAF during the Ogaden War. (EtAF)

A raging hatred of the Ethiopians was what motivated most SNA soldiers and WSLF insurgents (seen here) to keep on assaulting even the best protected and heavily fortified positions. (Albert Grandolini)

would not put these precious aircraft at risk for an operation that was not directly related to fighting the war in Ogaden, but to the Soviet wish for revenge. Theoretically, the EtAF could have flown that mission with its MiG-21bis's instead, but the Ethiopians concluded that only their F-5Es possessed the necessary range in combination with useful bombload. Namely, with all of its major bases constructed at a relatively high altitude above sea level, the MiG-21s would not be able to take the required full loads of fuel and bombs, while their take off would require an exceptionally long runway. Finally, the EtAF raid on Berbera AB had already caused such a public outcry in the Arab world, that Egyptian President Sadat publicly announced his intention to provide aid for the 'aerial protection' of Somalia, and Addis Ababa was certainly not interested in creating itself additional enemies.

Perhaps more importantly, the Ethiopians already knew that the EtAF attack on Berbera AB had thrown the CCS completely off balance. Combat losses had already reduced the Somali Air Force to only a shade of its former strength, but this blow finished it as a credible power. At the start of January 1978, it was barely capable of putting a pair of MiG-21s or MiG-17s into the air, and never ventured anywhere near the battlefield again. The only exceptions to this rule were two CCS An-24 transports: equipped with NKPB-7 visors actually used for dropping paratroopers, these could also be armed with bombs, and were subsequently deployed for several nocturnal attacks on isolated Ethiopian positions. However, even such operations were curbed when one of them only narrowly avoided an interception by EtAF F-5Es.

This is not to say that the CCS did not try to change the situation. Its commanders were exercising strong pressure upon President Barre to obtain replacement aircraft. According to unconfirmed reports, Somalia eventually contracted twenty Pakistan Air Force officers to retrain the remaining CCS pilots and the influence of Pakistani advisers eventually resulted in Mogadishu opening negotiations with Beijing for the acquisition of Shenyang F-6 fighters (a Chinese copy of the MiG-19S, that had been in service with the Pakistan Air Force since the late 1960s). Although concluded with an order, these negotiations did not help improve the situation immediately, as even the Chinese could not deliver any F-6s to Somalia before 1979. The SNA thus went into the last battles of the Ogaden War protected from the highly effective EtAF only by its flaks and SA-7s.

Meanwhile, as the first elements of the Cuban ground forces began appearing on the battlefields of the Ogaden War in early January 1978, Cuban pilots appear to have flown intensive reconnaissance sorties, collecting detailed information about Somali positions. According to Russian sources, the SNA's air defences in Ogaden were still alert and very active, and thus the DAAFAR pilots frequently mimicked Ethiopian F-5Es and MiG-21s that flew CAPs, prompting SNA air defences to remove their camouflage and open fire. However, whether this indeed happened, and whether the Cubans flew no less than 120 missions, detecting 136 important targets with only four MiG-21Rs by the end of January 1978, as claimed by Russians, remains unclear. As described above, the Ethiopians insist that they did not receive any MiG-21Rs before the end of war. Furthermore, the Ethiopians recall only very few instances in which their F-5Es had flown combined operations with either Cuban-flown MiG-21MFs and Ethiopian-flown MiG-21bis's. On the contrary, they insist that while many EtAF operations during December 1977 and January 1978 were hampered by bad weather, all the necessary reconnaissance sorties were flown by the single RF-5A and by two Canberras, indicating that the Cubans neither ever ventured beyond Somali borders nor needed any kind of top cover from F-5Es, because it was well-known that the CCS would not challenge them in the air.[125]

Final Somali Offensive

The Somalis might not have had a perfect picture about the flow of Soviet arms and Cuban troops into Ethiopia, but they had realised that time was working against them, and that they would be facing ever increasing odds, the longer the fighting continued. Siad Barre's government gambled high and lost the diplomatic battle, but was not ready to accept a diplomatic settlement. Instead, it was now dependent on SNA commanders to do their best and secure positions reached on the battlefields of the Ogaden War.

Although it suffered significant losses during first six months of the war, the SNA still had 22 brigades of battle-proven and hardened troops, including one intact mechanised and four motorised brigades, and could call on at least 135 MBTs, 205 artillery pieces and 100 other armoured vehicles.[126] According to Aidid, the major problem for these units was that Barre had left the entire SNDF without giving alternative plans for further operations and, at least as important, without supplies. Aidid later reported that he and the other SNA commanders repeatedly complained that neither Barre nor Maj Gen Gallel knew how much longer the fighting would go on, nor that most of the SNA's heavy weapons were lost or spent, and that they were in urgent need of reinforcements. Furthermore, severe pressure by the EtAF completely disrupted the flow of food and ammunition to the battlefield, and these were meanwhile in critically short supply. Not receiving their salaries for weeks and months, hungry and desperate, some of the Somali officers and NCOs then began stealing weapons and exchanging them for food with insurgents and local civilians.[127] Attempting to remedy the situation, SNA commanders ordered their units to conserve whatever supplies were left, while digging trenches, constructing fortifications and lying minefields, in order to secure their positions. Meanwhile, the remaining mobile units were prepared for one final offensive on Harar.

Exploiting a period of particularly bad weather that held most of the EtAF on the ground, the Somalis deployed nearly all of their remaining T-54 MBTs in support of a pincer attack by four infantry brigades against Fedis and Kombolcha, on 22 January 1978. The Ethiopian response came as a rude surprise. The EtAF fighter-bombers and Canberras not only hit additional supply depots in the rear, but also decimated several convoys moving in clear weather behind the battlefield, while assaulting Somali units were stopped by devastating defensive fire. Suffering, according to Ethiopian reports, as many as 3,000 casualties, the Somalis fell back.[128] A few hours later the Ethiopians launched their decisive counter offensive.

125 Ashenafi and Berhanu; ironically, in *With Foreigners against Our Own*, Babich in one place emphasises that MiG-21Rs were flown by 'Soviet pilots only', and in others praised Cuban pilots for their achievements with the aircraft.
126 Aidid et al., Chapter 14.
127 Ibid.
128 Ibid. Notable is that Aidid was convinced that his troops would have succeeded if he had been granted 'another try'. He put the blame on Maj Gen Gallel for not giving him the corresponding permission and ordering SNA forces to withdraw from Harar.

CHAPTER 6
THE ROUT

The Legend of 'Petrov's' Offensive

Most of the published accounts about the Ogaden War emphasise the importance of Soviet Lt Gen Petrov as the crucial commander during the offensive that ultimately forced the Somali military out of Ethiopia and ended the war. According to Ethiopian and even some Cuban accounts, the reality was entirely different.

The Derg began setting up their major counter offensive against the Somalis in Ogaden in late December 1977. In order to coordinate operations of Cuban, Ethiopian and South Yemeni contingents, the Supreme Military Strategic Committee (SMSC) was established in Dire Dawa, which consisted of top Ethiopian, Cuban and Soviet officers, and had the task of planning and running the coming offensive. This body did not function without quite a few complications. As the most senior member of the body and combat-hardened veteran of WWII, Deputy C-in-C Soviet Army, personal envoy of Soviet leader Brezhniev, and Chief Military Adviser to Mengistu, Lt Gen Petrov insisted on being appointed as top commander of the SMSC. However, he was not accepted by all the other members of this body, including Chief of Staff EA Maj Gen Haile Tilhaun, the C-in-C Southern Command EA Col Negash Mulatu, the new C-in-C EtAF Col Fanta Belay, and even the commander of the Cuban contingent, Gen Ochoa. Although there is little doubt that Petrov was considered a highly capable officer, the 'others' also found him very arrogant and unable to find common ground with Negash and Fanta in particular. According to Ethiopian sources, the main reason for this was the fact that Petrov was mainly

Colonel Fanta Belay took over as the C-in-C EtAF shortly after the Somali invasion. His aggressive and skilful deployment of the air force resulted in the destruction of the Somali Army's capability to complete the occupation of Ogaden. (EtAF via S.N.)

interested in removing all the remaining high-ranking Ethiopian officers trained in the USA. The fact that Negash and Fanta were not only US-trained, but also wore old imperial uniforms, similar in appearance to those of the corresponding US services, certainly did not help improve their relations with Petrov. The two Ethiopians repeatedly and openly disregarded his suggestions, usually making Petrov, who was certainly not used to any kind of contradiction, furious. Petrov's differences with other commanders reached such proportions that during the coming operation he began passing

Mengistu and his aides, such as Gen Haile Tilahun (a former IEAF F-5A pilot), were fast in understanding the necessity of demonstrating their 'socialist' political orientation for the purpose of attracting Soviet aid. By the time this Hungarian delegation visited them in Addis Ababa, pictures of Lenin and Castro were already hanging from walls where pictures of Emperor Selassie and the Ethiopian countryside used to hang at earlier times. (Robert Szombati Collection)

orders directly to Ethiopian field commanders who, of course, ignored him, provoking even more anger. At one point in time, a furious Gen Haile Tilhaun flew in a Huey to Harar Academy to confront Petrov and explain to him that nobody in the Ethiopian Army is ready to take orders from him. Ochoa reportedly shared this standpoint with his Ethiopian colleagues and went to great lengths in order to avoid even meeting him. Ultimately, it was only Mengistu's decision to raise Negash and Fanta in rank that settled the situation for the time being. If nothing else, it delivered a very clear message to Petrov about who had the final word in Ethiopia.[129]

Except for differences between Petrov and all the other members of the SMSC, everybody ably played his role and, at least officially, Mengistu never had enough praise for the Soviets, as long as Ethiopia got from Moscow whatever it wanted.[130] Indeed, after enraging Petrov through promoting Negash and Fanta, Mengistu went a step further and found it prudent to praise Petrov publicly for creating the plan for the upcoming operation and overall Ethiopian counteroffensive, although all the planning and commanding was undertaken by Haile Tilahun and Ochoa.[131]

The SMSC's plan was relatively simple; with the major objective being to stop the Somali advance, Ethiopian militia units, complemented by regular troops, were to hold the area east of Harar and Dire Dawa. Meanwhile, Cuban mechanised units would hit the Somalis in the north, where they expected them, and Ethiopian units would launch a general attack following the principle of double-envelopment. For this purpose, the Cubans were to advance north of Dire Dawa, reach a high plateau above the town and then take a sharp turn towards the south and hit the main SNA concentration in the flank. The Ethiopians would then clear the Harar–Dire Dawa axis, and push the Somalis in the direction of Jijiga.

By pure accident, this offensive was unleashed on 22 January 1978, simultaneously with the last Somali attack on Harar, and it started with a massive artillery barrage and aerial attacks that broke the back of Aidid's forces that were assaulting Fedis and Kombolcha.

129 'Off the record' commentary by several interviewed Ethiopian officers. Ironically, while in Ethiopia, Petrov lived at the commander's house of the Harar Military Academy and worked with his staff at the local offices, all of which were constructed with US aid. An UH-1H helicopter of the Army Aviation was put at his disposal and ferried him around for inspections of the battlefield, and although fiercely anti-American, Petrov learned to appreciate the Huey and began referring to it as 'Cobra'. Indeed, even Cuban Gen Ochoa followed this pattern and began to prefer the Huey, because it was smaller and more versatile than the large Mi-8, which required a long approach for landing.

130 This remained the case even after the Soviets became involved in a coup against him, more than ten years later.

131 Indeed, even nowadays, most Russian, and also several US accounts, report this operation as 'Petrov's Offensive', only a few observing that the assault of Jijiga was launched along plans prepared by Ochoa, though 'in cooperation with Soviet officers' (for example, see Babich, *With Foreigners against Our Own*).

The F-5E serial number 426 was one of the last four Tiger IIs that survived the Ogaden War. This photograph shows traces of repairs on the left side of the front fuselage, probably indicating the plane suffered some sort of combat damage. (Robert Szombati Collection)

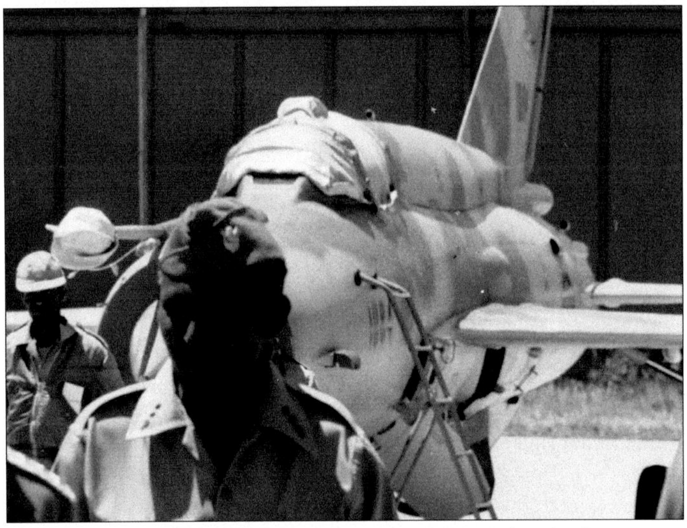

The MiG-21bis serial number 1084, seen here, belonged to the first batch of 48 such aircraft acquired by the EtAF. They bore the brunt of flying close-air-support sorties during the counteroffensive into Ogaden. (Robert Szombati Collection)

Furthermore, precisely because of the Somali attack on Harar, for the next 24 hours the SMSC was concerned that the CCS might attempt to interfere with EtAF and DAAFAR operations over Ogaden and therefore ordered a major aerial raid against Hargheisa AB.

Hargheisa – Revisited

The second EtAF raid on Hargheisa was much larger than the first one, the involved formations consisting of five elements, each assigned a different target. Ashenafi Gebre Tsadik led a flight (four aircraft) of F-5Es, providing top cover. Tesfu Desta led the first attacking element of four F-5As, while Heile Michael, Berhanu Wubneh and Wagira Gamechu led three sections of two MiG-21bis's each, respectively. The most important target of this attack, a radar station positioned in the hills east of Hargheisa AB, was assigned to the formation led by Berhanu. Basing their planning on intelligence provided by the Kosmos 964 satellite, the Soviets briefed Berhanu precisely about its location and even told him where to look for a dirt track leading to the site, for orientation. Deploying two bombs each, the MiGs knocked the radar out. Simultaneously, the F-5s and other MiG-21s cratered the runway and then hit other facilities, causing extensive damage. Once again, Somali interceptors were nowhere to be seen.

Meanwhile, additional formations of F-5As and the two remaining Canberras ranged far and wide over Ogaden, targeting the Somali supply system, as described above, while the DAAFAR contingent concentrated its operations on suppressing Somali air defences in preparation for coming operations. Their primary targets were the SNA's ZSU-23-4 Shilkas SPAAGs, but also stationary ZU-23-2s, M1939 flaks and even teams carrying SA-7 MANPADs. This was an exceptionally dangerous task indeed, and the first Cuban loss of this war was an EtAF MiG-21 flown by First Lt Raul Hernandez Vidal, shot down during one of such attacks, killing the pilot. Generally, Ethiopian and Cuban fighter-bombers operated separately from each other due to language barriers. During the following days, EtAF fighter-bombers therefore supported Ethiopian ground units, and Cubans their own.

On the ground, the operations developed relatively smoothly. The 11th Ethiopian Division, reinforced by one of two Cuban armoured brigades deployed in Ethiopia at the time (both were equipped with MBTs and APCs bought by the Ethiopians and underwent crash conversion courses to operate the equipment) advanced swiftly. By the morning of 23 January 1978, it regained all the territory as far as Fedis, capturing 15 Somali tanks, several APCs, as well as 48 artillery pieces and seven flaks in the process. The SMSC therefore issued an order to pursue without any respite and by 27 January, Somali units between Harar and Dire Dawa were practically enveloped.

Business end of a Somali 23mm ZU-23 flak. Positions of such weapons became the primary target of Cuban, but also Ethiopian, air strikes during the first days of the Ethiopian counter offensive. (Albert Grandolini Collection)

Two SNA T-54s and a collection of firearms captured by the Ethiopians and Cubans in the Harar area. (Albert Grandolini Collection)

Second Battle for Jijiga

On 1 February 1978, Ethiopian and Cuban ground forces launched a feint attack on Hawale, south of Dire Dawa, while the 1st Paracommando Brigade was deployed around the Somali flank with the help of helicopters and, when the SNA reacted with a counterattack on the following morning, attacked at Harewa from the rear. Taken completely by surprise, the Somalis hastily abandoned their positions. Three days later, as the Ethiopians and Cubans converged on Jijiga, the entire SNDF position was actually outflanked and in danger of being cut off from land connections to Hargheisa and Berbera. One of the SNA's brigades left to act as rear guard fell apart under pressure, leaving behind 42 tanks and even more APCs and other vehicles, many intact, and more than 50 artillery pieces. The SMSC then ordered the 69th Mechanised Militia Brigade and the 75th Militia Brigade into pursuit. In conjunction with the 1st Paracommando and 102nd Mechanised Brigades of the newly-deployed 10th Infantry Division that had surged from Kombolcha, these units eventually reached Jarso.

Meanwhile, the Ethiopians and Cubans continued their advances

Although overwhelmed by the force and firepower of their enemies, the Somalis fought bitterly until the last day of the war, and even managed to capture several Cuban soldiers. Three of these were photographed while being paraded down the streets of Mogadishu. (Albert Grandolini Collection)

aimed at re-capturing the Karamara Pass and attacking Jijiga. It was before this battle was joined that the next clash between Petrov and his counterparts within the SMSC occurred. Petrov demanded a frontal assault of Somali positions at Jijiga, accompanied by some flanking heliborne deployments of special troops. Arguing that such an operation against well-entrenched Somalis would prove very costly, Generals Fanta, Negash and Ochoa opposed this idea. Their reconnaissance had shown deep minefields in front of Somali positions, covered by a significant concentration of Somali artillery. Instead, they opted to tie down Somali units with the help of artillery and feint attacks, while the 10th Ethiopian Division with a subordinated Cuban mechanised brigade under the command of Gen Fleitas, circled the mountains south of Jijiga, while the other Cuban brigade, under the command of Gen Frias, crossed the mountains to the north-west of the town and thus closed the trap. Once again, Petrov was outvoted.

On the morning of 15 February 1978, the northern Ethiopian–Cuban task force bypassed the Karamara Pass and advanced on Arabi, proceeding to the Shebele Pass, more than 50km further north. Establishing a supply base at that site, this force then advanced on Grikocher and captured it thirteen days later, clearing the way for the newly-deployed 69th Militia Brigade, which then assaulted Lewenaji. Although suffering very heavy casualties – barely 500 of its soldiers survived the battle – the 69th Brigade then moved around Jijiga and captured the Kebribeyah–Aroresa road, and beat back a furious Somali counterattack destroying fourteen MBTs in the process.

The EtAF and DAAFAR supported this advance with all available means, setting the stage for the final assault on the strategic town. In the course of several dozens of sorties they hit concentrations of Somali artillery and different vehicle convoys. Resistance was fierce and a number of aircraft received hits from ground fire. Both Ethiopian Canberra bombers were damaged by ground fire and one of them was written off after returning to Debre Zeit AB, while the Cubans suffered two additional losses. The MiG-17 flown by Lt Eladio Campos was shot down and the pilot killed, while the MiG-21 flown by Col Benigno Gonzales Cortez was hit by ground fire but the pilot managed to eject safely after returning over Ethiopian Army positions. Furthermore, Capt Manuel Garcia lost the engine of his MiG-21 while flying at low altitude over Somali positions, but managed to start it again and return safely.[132]

With remnants of six SNA brigades and some minor WSLF-groups holding steadfast to their positions at Jijiga and only launching local counterattacks, the inevitable catastrophe occurred on 5 March 1978, when the SMSC initiated a large, combined-arms offensive. After more but 140 combat sorties flown by EtAF and DAAFAR F-5s and MiG-21s they softened up the Somali positions. The SMSC made use of 14th Squadron's Mi-8s to re-deploy the troops of the 1st Paracommando and 75th Brigades with stockpiles of ammo and fuel to a plateau north-east of Jijiga, where they joined the advancing Cuban armour and then attacked the enemy rear. The first Ethiopian units entered the town later the same day, though Somalis continued resisting until the evening of 6 March, when according to Aidid, all of their commanders received, 'a written but unsigned' order from Gen Gallel, ordering, 'a general retreat from the captured western Somaliland'. Supposedly issued by radio, during a meeting between Barre and Defence Minister Samater in Wardere, this order astounded most of the commanders in question, as recalled by Aidid:

> I had refused to accept such a humiliating, unwise and unsigned order. On receiving it, I immediately telexed to the President and Defence Minister, expressing my disagreement with the order and made my counter proposal for not withdrawing at this time. They approved it. I attacked the enemy forces with seven reinforced brigades. We had captured the railway. We had been holding our strong position between Dire Dawa (where the railway had been blown up) and keeping this with all our might, lest the Ethiopians should attack us. My forces were also holding the region of Hedogalo near the border. We distracted the enemy at the rear. The Russian experts had known me since the time they had worked with me in Somalia. They knew that I was at the front and so they changed their idea. I delayed their attack.

There is no evidence of the SNA managing to put up any kind of significant resistance after the loss of around 3,000 troops and most of its heavy weapons at Jijiga. As soon as that town was secured, the Ethiopians continued their advance along three routes, despite facing an increasingly difficult supply situation, which caused losses due to exhaustion and even dehydration.[133]

Race to the Border

After learning about the rout at Jijiga, Barre immediately announced that all Somali troops would be withdrawn from Ethiopia. It was too late, because the Cubans and Ethiopians then launched an outright race to reach the Somali border before the remnants of the SNA could do so. By 8 March, the 3rd Cuban Tank Brigade captured Degehabur, about 200km south of Jijiga, while the 94th Brigade of the 8th Ethiopian Division (the former 5th Division) took Fik against bitter resistance and, later the same day, the 3rd Paracommando Brigade broke into Kebridehar. The 69th Brigade recaptured Kelafo on the afternoon of the 13 March, and three days later the 12th Ethiopian Division seized Imi and el-Kere. With a clear indication of the Somali withdrawal turning into the almost complete destruction of the Somali Army, on 23 March Addis Ababa declared that the 'last frontier post' had been regained, marking the official end of the Ogaden War.[134]

Ultimately, the massive Cuban–Ethiopian assault, combined with heavy aerial bombardment of Somali lines of communication, but also the supply bases in and around Hargheisa, caused the breakdown of the Somali Army in Ogaden. While there was no doubt that the SNA officers and other ranks fought bravely and with skill, they were left with very little armour, no cover and were almost out of ammunition and supplies. As one military expert close to the Ethiopian High Command explained to the foreign media, 'They were sitting ducks. They didn't have a chance.'[135]

By that time, the EtAF was reinforced by the arrival of the first out of 44 MiG-23BNs ordered from the USSR, together with a group of Ethiopian and Cuban pilots freshly converted to that type.[136] These swing-wing fighter-bombers were rushed into action during early April, in the course of Operation Lash, which was mainly a counterinsurgency effort aimed at neutralising the WSLF in north-eastern Ogaden. The Somalis later reported a number of air combats and to have shot down 'several F-5s and MiG-21s', between 1 and 4 April.[137] In fact, two EtAF MiG-23BNs were shot down during Lash, and both of their pilots, Tadelle Alemu and Tadesse Mulluneh, ejected safely. Tadelle was captured and joined Lagesse Teferra in a Somali prison, to be exchanged in 1988, while Tadesse was recovered by an EtAF Mi-8.

However, Operation Lash was not really related to the Ogaden War as such. That conflict ended when the SMSC stopped the pursuit of scattered remnants of SNA and SALF units along the road to Berbera, on 7th April 1978, despite some reservations by Soviet advisers, and even Cubans, for continuing the offensive into Somalia.

132 Cooper et al., *African MiGs Vol. 1*, pp.160–169.

133 Tareke, p. 660 & Aidid et al., Chapter 14.

134 Tareke, p. 660.

135 Hussein, p. 47.

136 During the following years, the EtAF purchased more around 100 additional MiG-23BN fighter-bombers, MiG-23ML interceptors and MiG-23UB two-seat conversion trainers; for details see Cooper et al., *African MiGs Vol. 1*, pp. 160–169.

137 Kotlobovskiy, p. 32.

EPILOGUE

Although the Ogaden War came to an end in April 1978, the situation in the area, and especially along the Ethiopian–Somali border, remained tense for years to come. Reasons were multiple. Largely ignored by the Ethiopians and even the SNDF during the war, the WSLF was quite swift to recover and controlled much of the countryside again by September 1978. Under pressure after their defeat in the Ogaden War, which prompted a number of discontented SNA officers to launch an abortive coup against him, resulting with imprisonment or forceful emigration of most top CCS and SNA officers, Siad Barre attempted to exploit this development for his own purposes and began presenting the insurgency as 'fighting a Soviet-supported invasion', like in Afghanistan, to the Western media. Arguably, his efforts left most Western politicians unimpressed. Washington did enter an alliance with Somalia and the SNDF participated in some of the Bright Star exercises (in return, US pilots got to fly Somali MiG-21s), but refused to deliver any

Wreckage of the EtAF MiG-21bis serial number 1081 shot down by ground fire outside Tugwojale, in Somalia in July 1980. The aircraft skidded along the ground for more than a kilometre before the right undercarriage collapsed and the right wing embedded itself in the ground. The fire that broke out when the plane veered to the right side killed the pilot and completed the destruction. (Ground Defence International)

significant amount of armament, leaving it to Italy and various Arab states to help rebuild the CCS and SNA.

After spending most of 1978 and 1979 stabilising the situation in Eritrea, the Ethiopian military returned in force to Ogaden in early 1980, once again supported by a Cuban contingent and even a few Soviet advisers. However, its large-scale COIN offensive was largely unsuccessful and came to a sudden end in March of that year, when the WSLF launched a counteroffensive. This prompted the return of the EtAF to the scene again, and during July 1980 its fighter-bomber units flew dozens of strikes against insurgents, but also against targets up to 40km deep inside Somalia, losing one MiG-21bis in the process.

Following several additional offensives and counteroffensives, the large-scale warfare between Ethiopians and Somalis was mostly over by 1984, when the Ethiopians managed to secure most of Ogaden for themselves.

Price of War

There is still plenty of uncertainty over the cost of the Ogaden War, and most likely this question is going to remain unanswered, given the subsequent destruction of official Somali archives (and, indeed, the entire country).

Documents in the Ethiopian Ministry of National Defence cite figures varying from 5,137 to 6,650 officers, NCOs and other ranks killed, of which 160 were executed for a variety of reasons, ranging from cowardice to desertion attempts. Another 3,799 military persons are cited as missing in action, while 1,362 were categorised as deserters. The number of Soviet, Yemeni, and Cubans killed, wounded, or missing in action remains unknown, but is believed to be about 500 in total.[138]

Somali losses are much harder to quantify. Official Somali figures remain unavailable, while the Ethiopian Ministry of National Defence cited 6,453 Somali officers and other ranks as killed in action, 2,409 wounded, and between 109 and 275 captured. Most likely, these figures include losses the Ethiopians were able to positively confirm, primarily sustained during the second phase of the war. Russian sources confirmed that the Ogaden War 'weakened the Somali military significantly', and caused losses as high as 'one-third of the regular Somali National Army soldiers, three eighths of the armoured units and half of the Somali Air Force'.[139]

The results of the air war are at least as hard to summarise. During and immediately after the war, the CCS, SNA and WSLF claimed destruction of no less than 50 EtAF aircraft, primarily F-5s and MiG-21s, including about a dozen in air combat. With hindsight, it is certain that such figures are massively exaggerated. Any significant loss of Ethiopian aircraft would have almost destroyed the EtAF's chances of establishing air superiority over Ogaden, in turn meaning that the Somalis could have continued their offensive on Harar and beyond, and perhaps even that the Cubans and Soviets would have to have fought right after their arrival in Addis Ababa. This was not the case, and even Russian sources confirm that the Ethiopians established air superiority over the battlefield long before the Soviet air bridge of November 1977.[140]

The Ethiopian MoD cites a total loss of 23 EtAF aircraft, including two F-5As and three F-5Es shot down by ground fire, and one F-5E written off after suffering critical damage during the WSLF attack on Gode before the war. Berhanu Kebede was the pilot of both F-5As that were hit by ground fire, and he ejected safely in both cases. Afework Kidanu ejected safely from the first F-5E that was lost (at an as of yet unknown date), but was captured by Somali civilians and murdered. The loss of Lagesse Teferra and his F-5E was described above. The third Tiger II that was shot down is known to have been flown by Bacha Hunde and was hit by ground fire. Bacha made a crash landing and was picked up by an UH-1 army helicopter. Furthermore, one Canberra was written off during the

138 Tareke, p. 664.

139 Mezentzev, *Ethiopian–Somali Border War*, p. 36.

140 Babich, *With Foreigners against Our Own* & Kotlobovskiy, p. 32.

Wreckage of the F-5A 665 shot down during the Ogaden War under as of yet unknown circumstances. (via Pit Weinert)

war, after suffering a mechanical malfunction while returning from a combat sortie, perhaps also damaged by ground fire. The crew ejected safely near Debre Zeit AB. Sadly, like in so many cases, the exact date and detailed circumstances of that loss remain unknown. The other aircraft lost by the EtAF include one T-28 and eight SAAB B.17s destroyed on the ground at Dire Dawa AB when this was briefly captured by Somali troops, and up to three C-47s and (EAL-owned) DC-3s shot down or destroyed on the ground. While it seems that the EtAF did not lose any of its Ethiopian-flown MiG-21s during the war, two MiG-23BNs were shot down immediately afterwards. All of these losses occurred in the course of a total of 2,865 combat sorties flown by the Ethiopian Air Force between July 1977 and June 1978.

According to official Cuban sources, the DAAFAR pilots deployed in Ethiopia are said to have flown no less than 1,013 combat sorties between 22 January and 13 March 1978 alone, over 50% of them supposedly on MiG-17Fs. They suffered a loss of three aircraft and two pilots.[141]

The number of combat sorties flown by the CCS and precise details about its losses remain unknown. According to the Ethiopian MoD, the EtAF destroyed nine MiG-17s and eighteen MiG-21s in air combat, and as many as six 'MiGs' during attacks on Somali air bases. However, the EtAF officially credited its pilots with only eleven kills. For comparison, Russian sources credit the EtAF's F-5s with twelve Somali MiGs, and cite twelve or thirteen others as shot down by ground defences or crashed due to technical breakdowns.[142] While slightly more than a dozen Somali MiG-17s survived the war, according to recollections of former CCS pilots, only eight MiG-21s were left intact. Due to very low stocks of spares this type was subsequently withdrawn from service and replaced by Chinese-built F-6s in 1979.[143]

141 Cooper et al., *African MiGs Vol. 1*, p. 189, based on interview with Col Eduardo Gonzalez, retired DAAFAR MiG-21 and MiG-23 pilot, in June 2010.

142 Tareke, p. 665; Babich, *With Foreigners against Our Own* & Kotlobovskiy, p. 32.

143 Such recollections of former CCS MiG-21 pilots are countered by recollections of retired officers of US Air Force's 4477th Tactical Evaluation Flight 'Red Eagles', that visited Somalia in 1981 and flew several sorties with remaining MiG-21MFs and MiG-21UM's of the CCS. Relevant information came from Steve Davies, author of the book *Red Eagles: America's Secret MiGs*.

Reasons for Somali Defeat

As described above, from the standpoint of Ethiopian and Somali airmen involved in this conflict, and as confirmed by information collected by Ethiopian military intelligence, there is no doubt that it was the Ethiopian Air Force that won the battle for air superiority over the battlefield in the first months of the Ogaden War. There is also little doubt that the EtAF then exploited its control of the air to disrupt the Somali supply chain so successfully that the SNA's advance ran out of steam and the Somalis required months to build-up their stocks of supplies to a level where only limited offensive operations were possible, between October 1977 and January 1978. With this, the EtAF bought time for the political leadership in Addis Ababa to win the war on the diplomatic front. It was in the trenches of that diplomatic frontline, where the decision about the final outcome of the war was to fall. Here it is clear that in their enthusiasm, the Somali leaders grossly underestimated the ability of the Derg to manoeuvre for advantage, of Moscow's interest in gaining influence in Ethiopia, and the CPSU's wish to punish Somalia for the 'kick out'.

Complaints about incompetence of Somali military leadership appear to be only partially valid. When starting the war, Barre and his generals certainly never expected what was more or less a 'direct' Soviet intervention. They were aware of the SNDF's and their country's meagre resources and inability to fight a protracted war of attrition, as well as Ethiopian numerical superiority. They carefully selected the time of the invasion and launched a one-off, all-out effort that had a reasonable chance of success. Except in regards of their failure to understand the importance of air superiority and their gross underestimation of the EtAF, their planning and conduct of operations was exemplary. Critique that the Somali Army did not advance fast enough is rather not completely correct. During the first weeks of the campaign, the SNA units had to overcome not only very problematic terrain but also the bitter resistance of isolated Ethiopian garrisons. Subsequently, the Somalis had their units exposed to the full weight of fierce EtAF air attacks that left scores of vehicles, especially those carrying supplies, destroyed. Under such conditions, no faster advance was possible. The outcome of any battle pitting mechanised units exposed in open desert against air power should be well-known, at least since British–German clashes in Egypt and Libya during World War II.

While Aidid later argued that Barre, Samater and other top commanders made a mistake by ordering a hurried withdrawal of all Somali forces from Ogaden without the necessary planning and coordination, there is meanwhile little doubt that the leadership of the SNA was left without any choice but to attempt to save what was left of the SNA as a fighting force.

From an African-centric point of view, although there is little doubt that external intervention decisively shifted the balance of forces and decided the final outcome of the war, with the minor exceptions of Cuban and Yemeni involvement, it is important to observe that virtually all fighting was done by Ethiopians and Somalis. Combatants on both sides were well-motivated, fought with grim determination, boldly, fiercely and skilfully. Especially the Somali courage and readiness to self-sacrifice in spite of bitter resistance, simmering heat, problematic terrain and lack of supplies that would be hard to endure by any contemporary or even modern-day African armies. However, they were outmatched by skilful deployment of Ethiopian firepower, based on years of superior training. The Somalis lost the final battles not only due to sheer weight of numbers or the material superiority of their opponents, but foremost because of the skilful manoeuvring and

ተ.ቁ	ቀን	አውሮፕላን	የመታው ተዋጊ	የኪደ መፊቻ
1	24/07/77	ሚግ – 21	BZ	በዛብህ ጴጥሮስ
2	25/07/77	ሚግ –21	LG	ነገሠ ተፈራ
3	25/07/77	" – 17	LG	ለገሠ ተፈራ
4	25/07/77	" – 17	LG	ለገሠ ተፈራ
5	26/07/77	" – 21	LG,BZ	ነገሠ እና በዛብህ
6	29/07/77	" – 21	BT	ባጫ ሁንዴ
7	17/08/77	" – 21	BZ	በዛብህ
8	19/08/77	" – 21	AF	አፈወርቅ
9	21/08/77	" – 21	FI	አሸናፊ
10	01/09/77	" – 21	FI	አሸናፊ
11	01/09/77	" – 21	LG	ለገሠ

Scan from an EtAF document, listing officially confirmed and credited kills scored by Ethiopian pilots during the Ogaden War. It states the date of the kill, type of aircraft shot down and name of the scoring pilot. Two-letter initials mean the following: AF = Afework Kidanu (1 kill); BT = Bacha Hunde (1 kill); BZ = Bezabih Petros (2 ½ kills); FI = Ashenafi Gebre Tsadik (2 kills); and LG = Lagesse Teferra (4 ½ kills). (EtAF)

rapid advances of the Ethiopian Army. Unsurprisingly, some of the Ethiopian participants later observed that, 'if one would combine the Ethiopian Air Force with the Somali Army, they would result in Africa's dream military'.[144]

The F-5 vs MiG-21 Investigation

The crucial failure of Somali air power in the Ogaden War is more than obvious. Undeniably, the CCS fought as courageously as Somali ground units. Its pilots did not stop attempting to win the battle for air superiority until later, when they were simply left without aircraft to fly. While they thus cannot be blamed for not trying hard enough, it is their commanding officers who were responsible for not providing them with better training, armament and intelligence. Instead, the commanders in question grossly overestimated their own, and completely underestimated the abilities of the Ethiopians. Worse yet, they thus underestimated the impact of their own losses upon the course of the entire war.

The reason for the CCS losing the battle for air superiority might not be as obvious. With hindsight it is certain that the outcome was a result of superior training provided to EtAF pilots long before the war. However, considering results of pre and post-war comparisons between the F-5E and the MiG-21MF, it is easy to conclude that the superior training of Ethiopian pilots was supported by the superior technology in their hands, and that it is rather surprising that several of the involved parties preferred to ignore the obvious.

Namely, the CCS pilots flew very little before the war, and received next to no air combat training; indeed, most Somali MiG-21 pilots possessed no previous experience on other fighter jets but converted to the type straight out of their Flying School. The Soviets also made the mistake of not prompting the Somalis to take the Ethiopian F-5Es more seriously. Worse yet, even once they had sided with the Ethiopians, the Soviets proved anything but keen to put the Tiger II to a serious test. Instead, it was the Ethiopians and Cubans that, several months after the war, agreed to stage a series of sessions in dissimilar air combat manoeuvring, pitting EtAF F-5Es versus Cuban-flown MiG-21s and MiG-17s. The leading Ethiopian pilot involved was Lt Col Techane Mesfin, while the two Cubans selected for this exercise were Lt Col Jose Febles and Lt Col Luis Quiñones.[145]

Although granted the advantage of starting each training session from a favourable position, both Cubans 'lost' every simulated air combat, and badly. Expecting their beloved MiG-21MFs to prove better than the F-5E, Febles and Quiñones were amazed by this result, but realistic enough to conclude that, while superior at high altitudes and speeds, the MiG-21 was actually no match for the F-5E at slow speeds and low altitudes. However, once this exercise became known to the Soviets, it caused a massive uproar amongst their top advisers in Ethiopia. Petrov and Donikov, who considered it completely unacceptable that 'their' MiGs had lost the air war, and then simulated air combats against an air force

144 Tareke, p. 661.

145 Interviews with Ashenafi and Berhanu. Quiñones is known to have flown a high number of sorties during the war and later reached the rank of Brigadier-General of DAAFAR.

Major General Fanta Belay presiding over one of the post-war conferences, the aim of which was to study the lessons of the Ogaden War. To his right are Soviet and Cuban officers. Posters on the wall behind him are identifying aircraft and SAM-types in service with the EtAF. (EtAF via S.N.)

they were supposed to save, were enraged and went as far as to demand Techane be dismissed out of the EtAF on the spot. Only the personal intervention by Fanta Belay with Haile Mariam saved Techane's 'neck'.

Babich later explained:

> The Freedom Fighter's superior manoeuvrability to our MiG-17 proved a very unpleasant surprise. We received corresponding information only during the winter of 1978, when a Cuban unit equipped with this type arrived in Ethiopia. This was confirmed beyond any doubt during our testing. Surely, the MiG-17 was lighter and possessed an almost two times better wing to weight ratio in combat configuration, but its thrust to weight ratio was poorer. We expected that the MiG would outturn the F-5 easily but in reality one couldn't even dream about this! This American aircraft possessed a very advanced combination of flaps and slats, which eased the pilot's workload and aircraft's manoeuvring at high angle of attack. Important was also its thin wing, which offered better manoeuvring capability at slow speeds, or high speeds and high bank angle ... Overall, we were forced to realise that the 'Yankees' learned a lot from Vietnam. And still, we did not have the slightest idea about what was to happen only a few years later over the Beka'a Valley in Lebanon.[146]

Indeed, it took even the Ethiopians, who were preoccupied training additional pilots on MiG-21s and MiG-23s and with fighting the war in Eritrea, several years to draw all the necessary lessons from this exercise. Gradually, they realised that the CCS pilots they faced over Ogaden were too poorly trained in air combat. Sometime around 1981, when there was enough time, the EtAF organised several conferences, in the course of which it was concluded that the Soviets were providing only sub-standard training to their foreign clients. The Soviet training syllabus was based on Soviet doctrine and organised within a heavily controlled environment that required extensive ground facilities unsuitable to African conditions. In the air combat system deployed by the V-VS of the time, a highly experienced pilot sat in a control centre on the ground and controlled the entire battle by 'remote control'. Pilots flying aircraft were either fed instructions via radio or to their displays, telling them at what flight level to operate, which bearing to target and what power to apply to their engines. They were not trained to operate independently from outside support. Subjects like complex aerial operations, battle formations, and low-level navigation were not part of the curriculum. All of this was practically an anathema to

146 Babich, *With Foreigners against Our Own*.

Major General Fanta Belay with members of a Hungarian Air Force delegation visiting Debre Zeit AB in 1979. Interestingly, although officially 'visiting their allies in Africa', the Hungarians clandestinely took a number of photographs of various aircraft and installations at the main Ethiopian air base, many of which can be found in this book. (Robert Szombati Collection)

the Ethiopians, who based their operations on decision-making by the pilots themselves, low-level flying and deployment of decoys at every opportunity.

The EtAF stopped most of its own jet training in early 1974, and most of its pilots sent to train on T-37s in the USA were asked to leave in 1977. Still, this did not decrease the combat readiness of those pilots that already qualified on F-5s, particularly since by that time they had benefited massively from training in the USA. In comparison, the Soviets would not provide their Somali students with any kind of comparative training, or even the technical knowledge required for the SAC to become entirely self-sufficient. Although providing some assessments of the F-5, these were massively coloured by patriotism and often outright wrong. The EtAF might have been 'trained by imperialists', as the Soviets tended to describe it, but this training was not limited to the basics of flying and maintenance: the US training concepts also included combat and leadership training, and development of all support functions, including sporting and community events. This made the EtAF a closely tied community, an institution with its own culture and traditions where officers and pilots from different units knew each other intimately and thus cooperated in a much better fashion. The Ethiopian officers, who were spared no criticism or credit where deserved, eventually concluded that while anywhere else the Soviets went to in Africa they found newly independent nations that relied on them entirely for the creation of their air forces, the EtAF was a well-established air force with 30 years of uninterrupted development and experience with a well-developed doctrine. It could depend on operating technically superior aircraft, but also on the necessary technical, flying skills and experience not available in Somalia or similar countries. In the words of a former EtAF pilot that fought in the Ogaden War, 'No Soviet-trained air force stood a chance against an American-trained one'.[147]

Coda

Almost 40 years on, the Ogaden War continues to raise passions. The few pilots ready to talk about their experiences recount the terrible conditions in which they operated, the kills they scored and the comrades they lost. Although largely forgotten outside the continent, the Ogaden War is still casting a shadow over the Horn of Africa. Amongst the former battlefield, there are numerous remnants to remind people of the conflict, like desolate hulks of M41s and T-54s strewn around Ogaden, or faded photos and clippings recounting those heady days. Much more than this, the failure to capture Ogaden aggravated discontent with Barre's administration, eventually leading to its fall and then the dissolution of Somalia, resulting in the anarchy, chaos and protracted civil war, much of which is still going on today.

147 Former high-ranking EtAF officer, interview provided on condition of anonymity, 2008.

Bibliography

Aidid, (Gen) M. F., & Ruhela, Dr. Satya Pal., *Somalia: From the Dawn of Civilisation to the Modern Times* (Mogadishu, Vikas Publishing House, 1994)

Babich, Col Vladimir Kirillovitch, *Interceptors Changing Tactics* (*Istrebitely Menayut Taktiku*, in Russian), (Moscow, 1983)

Babich, Col Vladimir Kirillovitch., 'With Foreigners against Our Own' (in Russian), *Istoriya Aviyatsy* magazine, Volumes 25 and 29 (German translation kindly provided by Detlef Billig)

Berri, Capt M., *Aviation in Ethiopia* (*Aviation Be Ethiopia*, in Amharic), (Addis Ababa, Nigd Printing Press, 2002)

Brent, Winston, *African Air Forces* (Nelspruit: Freeworld Publications, 1999, ISBN 0-583880-6-7)

Cooper, T., Weinert, P., with Hinz, F., and Lepko, M., *African MiGs: MiGs and Sukhois in Service in Sub-Saharan Africa, Volume 2, Madagascar to Zimbabwe* (Houston, Harpia Publishing, 2011, ISBN 978-0-9825539-8-5)

Cooper, T., Nadew, S., Canyon, C., 'Fishbed gegen Tiger II: Luftkrieg über Ogaden', *Fliegerrevue Extra* magazine (in German) Volume 28/March 2010

Davies, Steve, *Red Eagles: America's Secret MiGs* (Oxford, Osprey Publishing, 2012, ISBN 978-1846039706)

Dupuy, Col T. N., Blanchard, Col W., *The Almanac of World Military Power* (2nd Edition) (London, Arthur Barker Ltd., 1972, ISBN 0-213-16418-3)

Editorial, 'The Non-War of Summer 1980 between Somalia and Ethiopia', *Ground Defence International*, No. 70, December 1980

Feleke, Zenebe, *It Happened Like That* (*Naber*, in Amharic), (printed privately, 2004); author was imprisoned from 1991 until 1998 and dedicated his publication to the person that helped him smuggle the original of this manuscript out of Ethiopia for publishing

Flintham, V., *Air Wars and Aircraft: a Detailed Record of Air Combat, 1945 to the Present* (London, Arms and Armour Press, 1989, ISBN 0-8536-779-X)

Forsgren, J., The Imperial Ethiopian Air Force, 1946–1961, *Small Air Forces Observer* (SAFO), Vol. 20/No.1, March 1996

Gilkes, P., 'Revolution and Military Strategy: The Ethiopian Army in the Ogaden and Eritrea, 1974–1984', Proceedings of the Eleventh International Conference of Ethiopian Studies (Addis Ababa, Institute of Ethiopian Studies, 1994)

Gorgu, Col Girma, *Let History Speak for Itself* (*Yinager Tariku*, in Amharic), (Miazia, Birhan and Selam Printing Press, 2003)

Green, W., & Fricker, J., *The Air Forces of the World: their History, Development and Present Strength* (London, MacDonald & Co Publishers, 1958)

Harrison, William A., 'Fairey Firefly Variants', *Wings of Fame* magazine, Volume 12 /April 1998

Hellström, Leif, 'Imperial Fireflies: the Long Life of Ethiopian Fairey Fireflies', *AirEnthusiast* magazine, No.122, March/April 2006

Hussein, Muhammed, *The History of Ogaden* (Fairfax, George Mason University, 1997)

Jackson, P., 'Death of an Air Force', *AirForces Monthly* magazine, June 1993

Jones, B., *English Electric Canberra and Martin B-57* (Wiltshire, the Crowood Press, 1999, ISBN 1-86126-255-8)

Kotlobovskiy, A. B., *MiG-21 in Local Conflicts* (*MiG-21 v Lokalnih Konfliktah*, in Russian), (Kiev, ArchivPress, 1997)

Marcus, Harold G., *A History of Ethiopia* (Updated Edition), (Berkeley, University of California Press, 1994, ISBN 0-520-22479-5)

Mezentzev, Dr Stanislav, 'Global Issues Affecting Africa: Ethiopian–Somali Border War of the 1977–1978 and Reflections on the Role and Impacts of Foreign Countries', *African Armed Forces Journal*, January 2013

Möller, Harald, DDR und Äthiopien, *Unterstützung für ein Militärregime (1977–1989)* (Berlin, Verlag Dr. Hans-Joachim Köster, 2003, ISBN 3-89574-492-1)

Nkaisserry, Brig Joseph K., *The Ogaden War: An Analysis of its Causes and its Impact on Regional Peace on the Horn of Africa* (Carlisle Barracks, Strategic Research Project, US Army War College, 1997)

Sarin., Gen O., Dvoretsky, Col L., *Allien Wars: The Soviet Union's Aggressions Against the World, 1919 to 1989* (Novato, Presidio Press, 1996)

Scuttts, J., *Northrop F-5/F-20* (Osceola: Motorbooks International, 1986, ISBN 0-7110-1576-7)

Tareke, Gebru, 'From Lash to Red Star: pitfalls of counter-insurgency in Ethiopia, 1980–82', *Journal of Modern African Studies*, Vol. 40/No. 3, 2002

Tareke, Gebru, 'The Ethiopia-Somalia War of 1977 Revisited', *The International Journal of African Historical Studies*, Vol. 33/No. 3, 2000

Turner, John W., *Continent Ablaze: the Insurgency Wars in Africa, 1960 to the Present* (London, Arms and Armour Press, 1998, ISBN 1-85409-128-X)

Urban, Mark, 'Soviet Intervention and the Ogaden Counteroffensive of 1978', *The RUSI Journal*, Volume 128, Issue 2, 1983

Wheeler, Barry C., *An Illustrated Guide to Aircraft Markings* (London, Salamander Books Ltd., 1986, ISBN 0-86101-206-2)

Interviews with various participants (see acknowledgments and footnotes); various articles in publications as mentioned in footnotes; *Newsweek* magazine (volumes from 27 June 1977, 29 August 1977, 12 September 1977, 28 November 1977, 13 February 1978, 20 February 1978, 13 March 1978) and *Insignia Magazine* Volume 6 (Blue Rider Publishing)

Acknowledgments

This book is the result of cooperation with a number of individuals from the Horn of Africa who generously helped with background and insider knowledge, relevant information and expertise; foremost of these were Major General Ashenafi Gebre Tsadik (EtAF retired), Colonel Berhanu Wubneh (EtAF retired) and Col Ahmad Sheikh (CCS retired). While establishing contact with them and prompting them to share their recollections proved a very complex and problematic task, there is little doubt that it would have been impossible to realise this project without their kind help and patience. Many others initially refused any kind of contact, but eventually agreed to grant verbal or written narratives solely on condition of absolute anonymity for reasons of concerns for their own and the safety of their families. I would like to express my special gratitude to everybody involved, but also to promise to keep the memory of a number of persons that have passed away in the meantime, or those that are in poor health now.

I would also like to thank to Albert Grandolini, Michel Benichou, and Jacques Guillem from France; Dr David Nicolle from Great Britain; Robert Szombati from Hungary; Pit Weinert from Germany; Jeroen Nijmeijer from the Netherlands; Mark Lepko, Tom Long and Adrian deVoe from the USA; Babak Taghvaee and Farzad Bishop from Iran; and Ugo Crisponi from Italy for providing extensive aid in photographic and other forms of research that eventually made this book possible; and to my wife, for always patiently accompanying me, at all times providing the support required to complete my research and 'put it to paper'.

Tom Cooper

Tom Cooper, from Austria, is a military-aviation journalist and historian. Following a career in a worldwide transportation business – in which, during his extensive travels in Europe and the Middle East, he established excellent contacts – he moved into writing. An earlier fascination with post-Second World War military aviation has narrowed to focus on smaller air forces and conflicts, about which he has collected extensive archives of material. Concentrating primarily on air warfare that has previously received scant attention, he specialises in investigative research on little-known African and Arab air forces, as well as the Iranian Air Force. Cooper has published 21 books – including the unique 'African MiGs' and 'Arab MiGs' series, which examine the deployment and service history of 23 African air forces during the period from the 1960s to the present day and major Arab air forces in conflicts with Israel, 1955–1973 – as well as over 200 articles on related topics, providing a window into a number of previously unexamined yet fascinating conflicts and relevant developments.